The
Carpenter's Bible

The Carpenter's Bible

A Home Owner's Bible

BRUCE CASSIDAY

DOUBLEDAY & COMPANY, INC., GARDEN CITY, NEW YORK

1981

Library of Congress Cataloging in Publication Data

Cassiday, Bruce.
  The carpenter's bible.

    1. Carpentry.   2. Dwellings—Maintenance and repair.
I. Title.
TH5607.C37   643'.7
ISBN: 0-385-11210-6
Library of Congress Catalog Card Number 77–82933
Copyright © 1981 by Bruce Cassiday
All Rights Reserved
Printed in the United States of America
First Edition

Contents

Introduction

Carpentry is the craft of making things out of wood. Carpentered objects can be as large as a building or as small as a toothpick. The most extensive use of carpentry has always been to provide housing for the human race.

Wood was one of the first construction materials adopted by man. The techniques and tools he created to shape wood to his will are used today with only slight modifications to work today's woods, metals, and plastics.

To keep an adequate roof over his head, the homeowner must be familiar with the elementary carpentry skills. Some of them are simple; others complicated. The fundamental facts about carpentry are covered in this book.

First, the amateur carpenter must have a basic knowledge of woodworking tools: what they are for and how they are used. These points are covered in Section I of this book.

Next, he should be familiar with wood and woodlike materials. This subject, including instructions on shopping for lumber, is covered in Section II.

Because repair work on the house constitutes a large percentage of the amateur carpenter's endeavors, he should know how to repair or replace damaged material. To understand how, he must know something about the principles of basic wood-frame-house construction. Section III contains detailed descriptions of interior construction and instructions on repair techniques. Section IV contains the same about exterior construction.

None of the projects or repairs in this book are beyond the capabilities of the average home handyman. With a little perseverance and adherence to the instructions and directions contained herein, anyone should be able to make his home or apartment a better, more solid place to live.

The Carpenter's Bible

Section I

TOOLS

CHAPTER ONE

Tools of Preparation

The first step in any woodworking job is to prepare the material for shaping by using cutting tools. Preparation of lumber is called "laying out." Laying out constitutes measuring, marking, and holding the material in place while it is being cut or shaped.

Proper measurement is an absolute must in laying out wood stock, as is the accuracy and straightness of the line marked for cutting. No carpentry job can be accurate if laid out wrong in the beginning.

Three important functions are performed by the various tools of preparation: (1) accurate measurement of length or distance; (2) accurate marking of right angles; and (3) installation of wood members in perfectly level positions, vertical and horizontal.

MEASURING LENGTH

Every carpentry job demands an accurate measuring tool to determine the proper length a piece of wood should be cut to fit in place. For lengths longer than 2 feet, you'll need either a folding wooden rule or a flexible steel tape.

The **folding wooden rule** is a convenient measure that can be extended from 4 to 8 feet without support. The rule is made up of a number of 6- or 8-inch wooden sections hinged together to fold out to a rigid length.

The rule is sharply angled to operate around corners, for measuring recesses or other blind entries, and to aid in transferring angles to wood stock for fitting boards into odd-shaped corners.

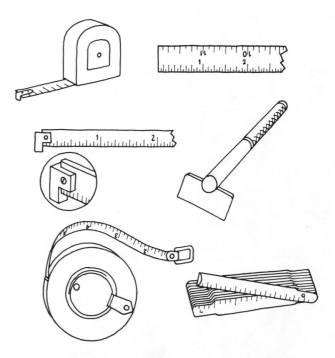

Fig. 1. The carpenter's folding wooden rule is used to make straight across-gap measurements, measurements around square corners, and interior measurements. The rule folds up into a compact unit when not in use. Picture shows several common types of rules, including the folding rule.

The rule can be used to measure along a wall to the ceiling, or to determine the distance to one particular spot from where you are standing. Some models have a sliding extension that makes it possible to determine precise interior measurements in places inaccessible to a straight ruler

longer than 6 or 8 inches. An interior measurement is any kind of distance to be determined *inside* a recess, like a closet or box where an ordinary yardstick or ruler will not fit.

The folding rule can also be used like any other type of rule to mark lengths on lumber and other construction material.

The **flexible steel tape** is compact and easy to use, with 6- to 12-foot lengths commonly available. Its flexible steel ribbon rolls up into a compact steel case 2 inches square. Because its base is 2 inches long, you can make interior measurements simply by adding 2 inches to the tape's visible reading.

The extended ribbon rolls up automatically into the case when you finish with it. Some models have a locking device to keep the tape in extended position while measuring. The inch designations are divided into 16-inch groups for quick-see marking of material for installation on wall studs placed exactly 16 inches apart.

MEASURING RIGHT ANGLES

Carpentry is basically the craft of fitting wood pieces together at right angles. Every box has bottom and sides; every house has walls and floors. When walls are out of parallel, the structure tends to fall. When a box is not properly built, it tends to break.

Because right-angle marking is crucial, the square is one of the most important layout tools in the carpenter's box. There are several types of squares that have been developed for carpentry, each designed for a particular type of work: the steel square; the try square; the combination square; and the try-miter square.

The **steel square,** also called the framing square, is a sturdy L-shaped steel measuring tool 12 inches wide and 18 inches long. Other sizes of $1^6/_{24}$, and $1^8/_{24}$ are also available, although the $1^2/_{18}$ is the most commonly used by the amateur carpenter. It is employed mainly for marking off right-angle cuts on straight pieces of lumber. You lay the long edge of the square along the side of the board and mark along the short edge. The square can also be used to check a right-angle fit after two boards are fastened together.

Fig. 2. Some flexible-steel tapes like this one roll all the way out to 25 feet, with an inch-wide blade sturdy enough to support itself. Most tapes come in a lightweight case with a clip for the belt and a thumb lock holding the blade firmly in place. Photo courtesy Stanley Corp.

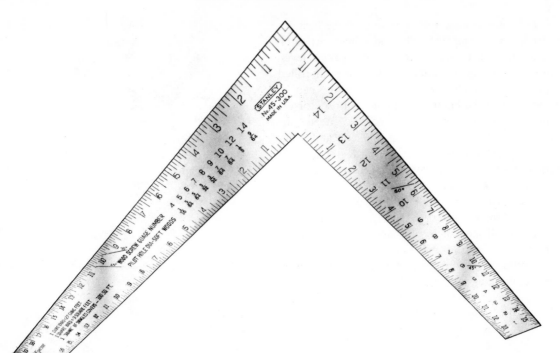

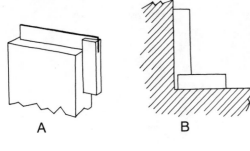

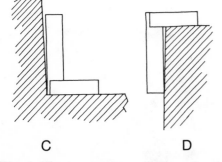

Fig. 3. This large-size steel square is 16 inches on the short side and 24 inches on the long, with the inches broken down into eighths. The square also includes a lumber scale showing board feet equivalent, a decimal table converting fractions to decimals, a metric conversion table, a formula for squaring a foundation, 45-, 60-, and 30-degree angle markings, volume and area formulas, a table showing the drill sizes for pilot holes of nine wood-screw gauge numbers, a table showing quantity per pound for various sizes of common and finish nails, and a depth scale. Photo courtesy Stanley Corp.

Fig. 4. The try square, as well as the steel square, can be used to check both inner and outer right angles as shown. In (A) try square is used to check right angle on sawed board. In (B) it is used to check inner angle. In (C) it is used to check a faulty inner right angle. In (D) it is used to check a faulty outer right angle.

The **try square** is an L-shaped marking tool with a steel blade along the long side and a wooden handle on the short side. Available in blade lengths of 3 to 15 inches, it is used for laying out right angles and especially for testing whether the side of a board is square to the surface.

To test a board's edge for squareness, place the handle of the try square firmly along one surface, sliding the blade into contact with the board's edge. If you can see light between the blade and the board, the edge is untrue. Sand or plane the edge until the try square proves it true.

The **combination square** is one of the best all-around measuring tools for the home carpenter. It is a try square, steel square, miter gauge, level, plumb, gauging tool, and 12-inch rule all rolled into one. A freely sliding 12-inch handle can be tightened to the blade or removed entirely for ordinary measurements. The handle meets the blade at a 45-degree angle on one side and a 90-degree angle on the other. The 45-degree angle is for use with miter angles—both to lay them out and to check them for accuracy when cut.

There are many different types of combination squares, some with such esoteric measurements as dowel diameters, lengths and shanks of common nails, lengths and gauges of wood screws, and other measurements. Most have level, bevel angle, marking tool, and a 12-inch rule.

B

C

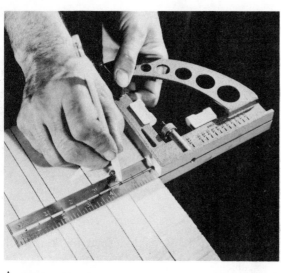

A

Fig. 5. This type of combination square can be used for scribing circles, for planning angled cuts, for checking squareness, plumb and level, and for determining common nail, wood screw, and dowel sizes. The combination is being used as a scribing tool in (A) to mark a rip cut on a board. (B) Tool being used to make arcs. (C) Tool being used for measurement of nail and screw shank sizes. (D) Tool being used to measure a true right angle. Photo courtesy Stanley Corp.

D

LEVELS AND PLUMBS

Carpentry is as much the art of determining the true vertical as it is of measuring lengths and marking right angles. Every structure that is erected must stand at right angles to the ground or it will tend to topple over. A board vertical to the ground has its center of gravity exactly in the middle and is less likely to fall.

Windows and doors are built with horizontal and vertical planes in mind. To get walls straight and floors level are crucial considerations that must be kept in mind at all times when repairing any portion of a house. For this work you can use the level and the plumb.

The **level** is a tool with a glass tube holding an air bubble in water or fluid. The level is maneuvered until the bubble moves between two marked lines; then the tool is level. A board in that position is also level. At one end of the level there is usually a vertical bubble tube to indicate plumb in a vertical plane.

The **plumb bob** can also be used to check for true vertical. It is simply a string with a lead weight attached to the end in the shape of an elongated cone. With the end of the string held in the air, and the weight at the end, the plumb bob will settle into position at true vertical. The plumb bob can be particularly useful in erecting a fence or wall.

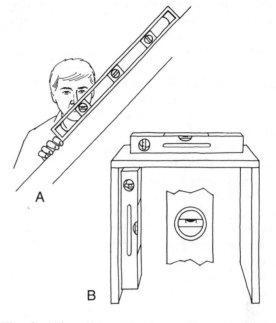

Fig. 6. Although most levels contain only vertical and horizontal measurements, the one in (A) checks a 45-degree angle as well. Bubble must be exactly in the middle of the tube for accurate readings, as shown in (B).

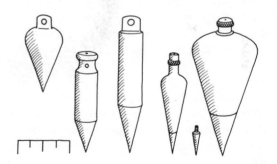

Fig. 7. A variety of plumb bobs, made of brass, steel, and combinations of both. The bob shown at left is the most common, and usually the simplest kind for the amateur carpenter.

MEASURING OTHER ANGLES

The **sliding T-bevel** is an angle-measuring tool usually used in combination with a square. It resembles a try square, but is adjustable to any angle. It is used most often to transfer a specific angle from one piece of stock to another, or to measure an angle and then transfer the angle to the stock for cutting. The steel blade has a 45-degree bevel on one end for convenience.

TOOLS THAT MARK

Even though the pencil is the most familiar marking device around the house and can be used in most cases by the amateur carpenter, there are other markers more accurate and more applicable in certain instances. These are the marking gauge, wing dividers, and the chalk line.

A **marking gauge** is the proper tool to use in marking a line parallel to a board or panel edge, either to prepare for ripping or for fitting another board to the panel. It consists of a thick rule mounted on a square strip of wood with a sharp pin at the end of the rule. Set the gauge, slide it along the wood to be ripped, and it will mark a cutting line along the edge of the wood (Fig. 8).

The **wing divider,** similar to an arithmetic-class compass, is used to mark circles and arcs on wood. Both legs of the steel divider have points, with one anchored in the center of the circle, and the other scribing the wood. The diameter can be set conveniently by a screw head.

The **chalk line** is used to mark a straight guide line on a surface, usually for the purpose of laying brick, or marking the boundaries of a square or rectangle on ground or floor. The line is simply a cord covered with chalk dust. With the cord stretched out straight, one end tied to a nail or stake at the far end of the line, pull the cord tight and slap the cord along the surface to be marked. The action flicks a straight guide line of chalk dust on the surface, which can be brushed away later.

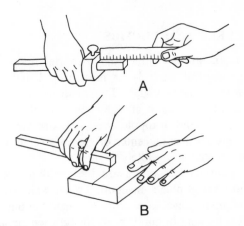

Fig. 8. The marking gauge can be used for accurate scribing to guide the ripsaw along a board. First measure the distance from the pin to the flat side of the wood strip, as in (A). Then use the gauge as a scriber, running it along the flat side of the board to be sawed, as in (B).

Fig. 9. Wing divider is used for making circles and arcs. Steel bar separates the legs, with a locking nut for securing rough measurements, and an adjusting screw for finely calibrated work.

TOOLS THAT CLAMP

The clamp is the carpenter's third and fourth hands—some of them even look like clawed hands—and is used not only for holding stock while it is being worked, but also for keeping two pieces together after gluing or applying adhesive.

A clamp can be destructive if it is screwed too tightly to a piece of finish wood. To keep from ruining expensive stock, insert a scrap strip of wood between the jaws of the clamp and the surface.

The biggest and most obvious clamp of all is used to hold stock while the carpenter works on it. This clamp is called a vise.

The **woodworker's vise** can be clamped onto your workbench or on a sawhorse or on any table where it will hold many materials securely, freeing both your hands for work with tools.

The best kind of vise to get is one with large jaws that will distribute pressure evenly. Most vises come without hardwood faces—you add the wood with small screws. When mounting a vise, keep the top of its jaws flush with the bench top.

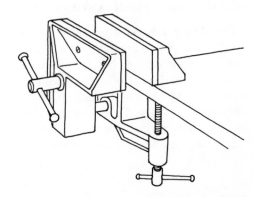

Fig. 10. Typical vise is composed of bench clamp, which secures it to any worktable, and the vise cinch, which closes the jaws to hold working material.

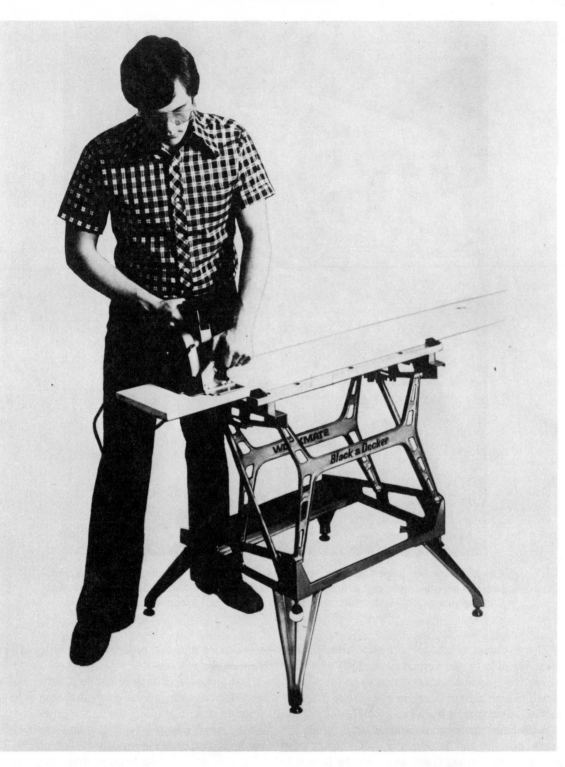

Fig. 11. Combination sawhorse and workbench makes it easy to cut and shape large pieces of lumber. Securing the material leaves both hands free to hold and guide the tool, while the wood stock remains firmly in place. Photo courtesy Black & Decker.

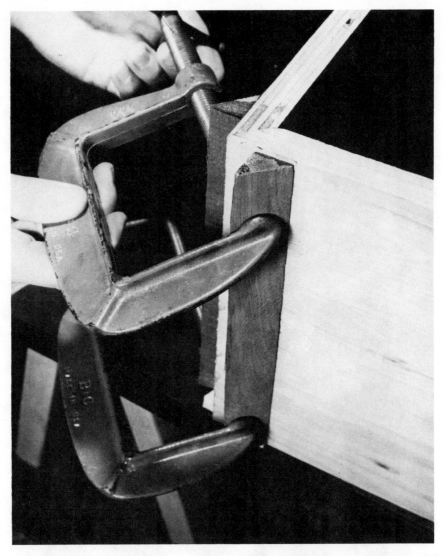

Fig. 12. Here two large C-clamps are used to hold together a miter joint for gluing. Outside triangular pieces have been attached to the two mitered members so that C-clamp pressure holds the miter together. Photo courtesy American Plywood Assoc.

The **C-clamp,** available in many sizes, is the most versatile of the carpentry clamps, and can be used to hold materials together or to the workbench. These clamps come in sizes from 3 to 16 inches. Some special types are smaller.

Adjustable hand-screw clamps are used to apply pressure on flat surfaces being glued together. Both the angle and the distance between the wooden jaws are adjustable, making irregular flat-sided objects easy to clamp. Be sure to keep the angles of the jaws parallel to the angle of the work being glued in order to maintain even pressure on flat surfaces.

Bar and pipe clamps resemble a lathe, and are used for clamping across a long expanse. The adjustable jaws slide along the bar or pipe.

The **band clamp** is versatile and can be used for holding together unusually shaped pieces. The clamp consists of a canvas-strap loop drawn tightly into a metal buckle, holding a strange shape rigid while glue dries or while it is being worked with a cutting or shaping tool.

Fig. 13. Large adjustable hand-screw clamps can be used with great effectiveness in holding together wooden pieces for gluing. The hand screws are easy to manipulate to attain the correct angle against the glued stock. Clamps can be used on irregularly sided objects as well. Jaws of clamps must be kept parallel to the work. Photo courtesy American Plywood Assoc.

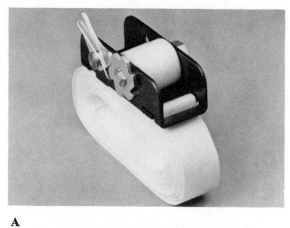

A

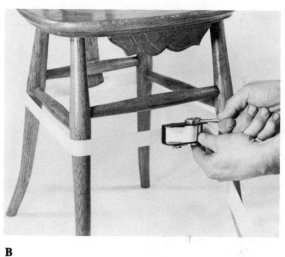

B

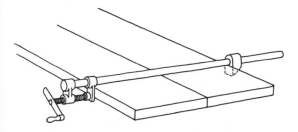

Fig. 14. Pipe clamp is used when very wide spans of material must be braced for gluing. Left side of pipe clamp contains adjustable cinch, and right side has jaw held tightly against board. Courtesy Stanley Corp.

Fig. 15. Flexible band clamp consists of a strap loop and strong metal buckle (A). For odd-shaped objects, as well as strictly right-angled ones, the band clamp can be used for comparatively wide distances that can't be spanned by a C-clamp or hand-screw clamps. Metal buckle is tightened with wrench (B). Photo courtesy Stanley Corp.

CHAPTER TWO

Shaping Tools

THE SAW

The saw is the most common shaping tool in the workshop, designed to cut either against the grain or with the grain of lumber. Cutting against the grain is called crosscutting, and cutting with the grain is called ripping. While most saws can be used for both ripping and for crosscutting, there are special saws made especially for each job.

The most important element in any saw blade is the number, size, shape, and direction of its teeth. Each tooth is "set" or angled alternately to left and right to give a cut wider than the blade itself to prevent the blade from binding in the wood. The set determines how wide the cut—or "kerf"—will be. The softer and less seasoned the wood, the wider the cut necessary.

Six types of saws are commonly used in the home workshop: crosscut saw, ripsaw, keyhole or compass saw, backsaw, coping saw, and hacksaw.

The **crosscut saw** is designed for cutting against the wood grain. Its teeth act as small parallel knives set apart to keep the kerf wider than the blade. The crosscut saw does most of its cutting on the downstroke, with very little on the upstroke. For thick boards, a crosscut saw with coarser teeth will cut faster than one with close teeth. A blade with 7 to 8 teeth to the inch is satisfactory for the home carpenter. If fine work is contemplated, 10 teeth to the inch will do a cleaner job. Crosscut saw blades vary in length from 20 to 28 inches. A 26-inch blade will perform most jobs around the house satisfactorily.

Before beginning a cut, secure the board to be trimmed either in a large vise or under your knee on a sawhorse or flat bench. If the board is long, place it at right angles to two sawhorses, and plan to cut to the right of the second horse.

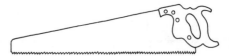

Fig. 16. Typical handsaw. The handle is usually made of ash or a similar hardwood, and is riveted to the end of the steel blade.

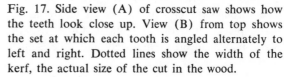

Fig. 17. Side view (A) of crosscut saw shows how the teeth look close up. View (B) from top shows the set at which each tooth is angled alternately to left and right. Dotted lines show the width of the kerf, the actual size of the cut in the wood.

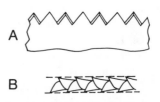

Fig. 18. Close-up of a crosscut saw's teeth, showing the degree of bevel (angle) of each along with alternating bevels.

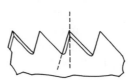

Fig. 19. Close-up drawing shows the teeth of an ordinary crosscut saw blade.

To begin a cut, grasp the handle of the saw firmly with the right hand, the thumb and index finger touching the side of the handle. (Left hand, of course, if you're left-handed.) Do not start the cut in the middle of your mark. Start on the waste side of the line, making sure the wood you want is on the good side. This way you will insure an even surface: by leaving the line intact you will indicate where the stock should be planed or sanded when finished.

Move the saw up and down several times, guiding the blade with the left-hand thumb knuckle. Draw the blade *up* slowly at the point where the cut is to be made. Do not push down to start the cut.

Once you have made a preliminary cut in this fashion, hold the saw perpendicular to the board and push down slowly along the cut, deepening it. If the saw jumps out of the groove, put it back in place.

When the blade is moving easily through the cut, attack the board at a 45-degree angle, and continue sawing with long, slow, easy strokes, moving the saw from the top down putting pressure on the downstroke, and drawing back easily on the upstroke. If the blade sticks, don't force it. Take short up-and-down strokes to loosen the blade and enlarge the kerf if there is resistance. At the end of the cut, shorten your strokes and be sure to provide support for the waste piece. Saw through the end of the cut as you hold the waste piece with your free hand. Otherwise the waste may snap off prematurely and take a piece of the finish wood with it.

If the saw leaves the line as you are working, twist the handle slightly to draw it back in line. If the saw is not squarely vertical, bend it a little and gradually straighten it to the line of cut. Do not bend or kink the blade.

If you are cutting difficult portions of a piece of wood with knots or thicker segments, guide the blade forcefully, using the stronger part of the saw near the handle.

The **ripsaw** is designed to cut wood with the grain. Because the ripsaw blade meets less resistance than a crosscut blade, its teeth are larger and fewer in number, usually 5½ to 6 points to the inch. The teeth are set at an almost 90-degree angle to the blade to rip as well as cut the wood

Fig. 20. Mount board on sawhorse for best cutting position. If you're right-handed, left knee should brace board against the sawhorse.

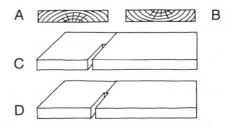

Fig. 21. When cutting across the grain with a crosscut blade, place the board to be sawed as shown in (A). This positioning will avoid splitting off the lower bottom of the board when you finish the cut, as would happen in (B). When making any cut, be sure the saw's kerf is on the bad side of the stock, not on the good. Make your cut as illustrated in (C), not in (D).

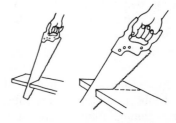

Fig. 22. In crosscutting, when the blade is moving steadily through the cut, attack the board at a 45-degree angle, sawing with long, easy strokes.

fibers. A thicker, coarser tooth is required for cutting thick stock; a finer tooth for thin stock. A ripsaw with a blade 26 inches long and with 5½ points is adequate for the home workshop.

When ripping wood with the grain, hold the saw at about 60 degrees to the plane of the wood. If the board to be ripped is a long piece, clamp a straight-edged length of wood on the cut line to guide your saw. On a long rip job, insert a wedge in the end of the cut to keep the sides from pressing in and jamming the blade.

The **keyhole** or **compass saw** is designed for cutting curves or holes. It has a narrow, tapered blade that can fit into narrow spaces to deliver twists and turns in any wood stock. A typical compass saw has a 12- to 14-inch blade, with 8 to 10 teeth to the inch. A keyhole saw usually has a narrow blade, and is 10 to 12 inches in length, with 10 teeth per inch. The keyhole saw cuts a smaller diameter than the compass saw.

Both keyhole and compass saws are used for cutting openings in floors or walls for pipes and electric outlets with the cut starting from a bored hole. They can also be used for making short cuts of any kind, to finish off a long cut, or to make irregular curlecues.

The **coping saw** is designed for curve cutting, fillagree work, and curlecues. It cuts curves smaller than ¼ inch. It has a replaceable blade of very thin metal varying in thickness from $\frac{7}{100}$-inch to ⅛-inch thick with 10 to 20 teeth per inch. Blades are available for cutting plastics as well as thin metal and wood. The blade is changed by loosening the threaded grip. The coping saw in varying sizes is called the jigsaw, the scroll saw, the fret saw, or deep-throat coping saw.

To make an interior cut, drill a hole and thread the loose blade through the hole. Then attach the blade to the saw's C-shaped frame. Most frames are designed so that the blade can turn in the frame during cutting. Depth between blade and back of frame is usually 4½ to 6½ inches.

For delicate, exacting scrollwork performed in a vise, the teeth of the blade should point toward the handle, with the cutting done on the pull stroke. For work supported on a sawhorse or bench, the points should turn away from the handle.

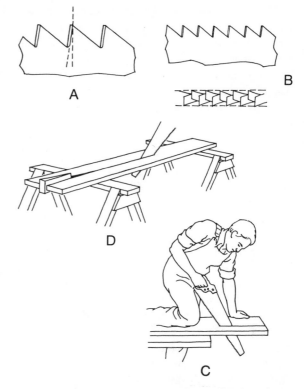

Fig. 23. Close-up (A) shows the teeth of a ripsaw blade, with the angle of the cutting edge visible. (B) The wide set of the teeth is shown in the inset, along with a typical ripsaw profile. (C) This is how to position board on sawhorse or worktable to make a ripsaw cut. After proceeding a number of inches, insert a wedge (D), to keep the pieces apart so the saw won't jam.

Fig. 24. The compass saw.

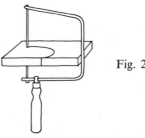

Fig. 25. The coping saw.

The **backsaw** has a reinforced back to provide rigidity in delivering a straight-line cut. It is designed for short, precise cuts, like miters and other joints. The backsaw is usually a crosscut, 10 to 16 inches in length, with 12 to 14 teeth per inch. It will cut with or against the grain.

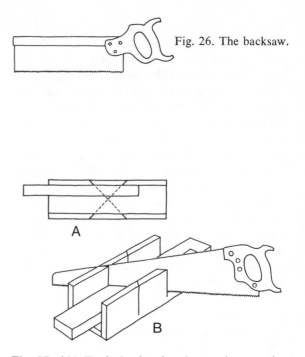

Fig. 26. The backsaw.

Although it can be used alone to make cuts, the backsaw is designed primarily for use in a miter box. A miter box is a wooden or metal frame in which angle cuts are made accurately. A wooden miter box has pilot slots for 45-degree angles in both directions, and for square-off cuts. The molding or wood strip to be mitered is placed in the box, and the backsaw is then fitted into the pilot slots to make the angle cut. A metal miter box can be adjusted to make cuts at any angle desired.

To use a backsaw in a miter box, mark the work for the cut, then line up the mark with the pilot slots. Make sure the cut is on the waste side of the line. Hold the work in the box and start with a back stroke, holding the handle end tilted slightly upward, leveling it as you continue.

If a miter box is not available, it is possible to make a miter cut with a backsaw alone. To start a backsaw cut, make a slight groove along the cut on the waste side, then check the accuracy of the angle with a combination square. Hold the backsaw along the entire length of the cutting line as you saw. Support the board being cut with an underlying piece of scrap wood so it will not splinter at the end of the cut. Do not use the backsaw to cut through nails, dry wall, or painted lumber.

The **miter-box saw** is simply a longer version of the backsaw, running from 20 to 26 inches, with 11 teeth per inch.

The **hacksaw** is designed for cutting metals. It has a changeable blade, with many variations intended for use with specific metals. The hacksaw's teeth point forward for cutting on the forward stroke.

A blade of 14 teeth per inch is designed to cut machine steel, cast iron, brass, copper, aluminum, or bronze more than 1 inch thick. A blade of 18 teeth per inch is designed for annealed steel, tool steel, high-speed steel, bronze, aluminum, and copper ¼ to 1 inch thick. A blade of 24 teeth per inch is designed for use with electrical conduit, drill rod, copper and brass tubing, wrought-iron pipe, steel, iron from ⅛ to ¼ inch

Fig. 27. (A) Typical miter box is seen in operation from the top, with a piece of wood in place for mitering. The angle of the cut is exactly 45 degrees to the length. (B) A handsaw is being used to cut a piece of wood in the box. Backsaw should be used for miter box work for best results.

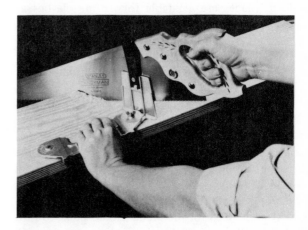

Fig. 28. This special saw-angle guide allows you to make square and miter cuts at angles of 45, 60, 75, and 90 degrees, without the use of a miter box. It can be used as shown with a regular crosscut saw. Photo courtesy Stanley Corp.

thick. A blade with 32 teeth is designed for use with these same materials ⅛ inch thick or less.

In carpentry the hacksaw is used for shortening screws or bolts when necessary, for cutting through nails in wood being sawed, and for cutting hardware or lengths of pipe.

When using a hacksaw, mount the metal in a vise so the work cannot shift and break the blade. In sawing, grip the handle in the right hand, and hold the frame in front with the left, making slow strokes with moderate downward pressure on the forward cutting stroke, and almost no pressure on the back stroke.

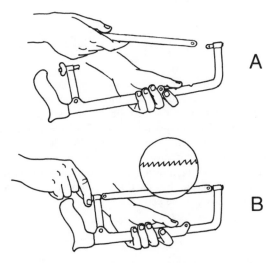

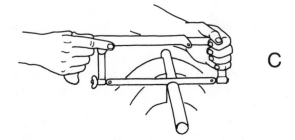

Fig. 29. (A) and (B) show how to place a hacksaw blade in a saw, with teeth pointing forward. (C) How to hold hacksaw when using it on stock clamped in a vise.

THE UTILITY KNIFE

There are a hundred uses for a utility knife in a home workshop: cutting Sheetrock, scoring resilient tile, working on trim, cutting shingles, and endless other small jobs.

When working with thin pieces of wood or trim, use the utility knife for making a line along the cut to guide the saw. In making joints, use the utility knife for cleaning areas where a chisel cannot be used.

The most satisfactory utility knife is one with a retractable blade and a button that prevents the blade from sliding when in use. The blade should be adjustable for two or more cutting positions so that a small cut or groove can be made without using the entire blade.

Some models can be converted into scrapers for removing paint or putty, and some have substitute saw blades. The knife blade should be of good steel and it should always be kept sharp.

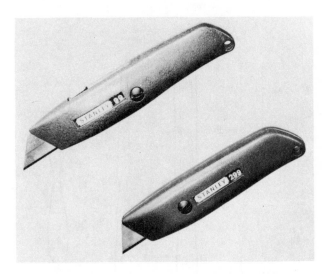

Fig. 30. The blade of a typical utility knife locks in three positions for different cutting depths. In the first position (not shown) the blade is partially extended for scoring, for opening cartons without damaging contents, and so on. The second position, bottom, is an intermediate position for special work. In the third position, top, the blade is fully extended for normal cutting jobs—wood, wallboard, rubber, and other materials. Photo courtesy Stanley Corp.

THE CHISEL

A chisel is a stronger version of the utility knife, used for cutting small grooves and notches in wood and for making miters, mortises, and other joints. Unlike the utility knife, the chisel has a bevel on the cutting edge.

The chisel consists of a handle and a steel blade. A wooden-handled chisel should be driven with the hand alone or with a wooden mallet; the metal-capped plastic-handled chisel, with a hammer.

Chisels come in widths of ⅛ to 2 inches. The smallest width is called a **mortise chisel,** the largest a **bevel-edge** chisel.

The **butt chisel** is the shortest at 7 to 9 inches in length, for use in confined spaces. The **firmer chisel** is 9 to 10½ inches long, and is used in general shop work. There are larger sizes, like the **mill chisel** at 16 inches, but the butt and firmer chisels are the only two likely to be used by the home carpenter.

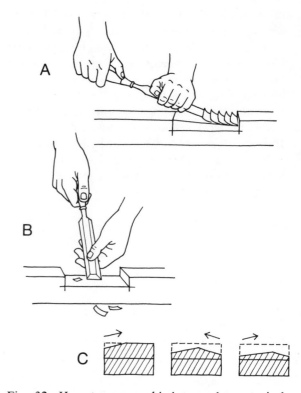

Fig. 32. How to use a chisel to make a typical notched cut, called a dado, in a piece of wood. Note (A and B) how the left hand guides the cutting, with the right supplying power. Use the bevel side down as in (A) for rough cutting. Use the bevel side up for smoothing out the bottom of the cut (B). (C) How to avoid splintering the corners when chiseling across the grain by starting at opposite sides and working toward the center.

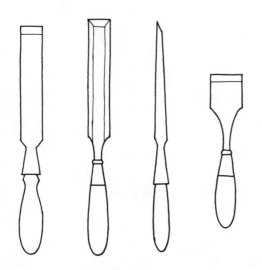

Fig. 31. Various types of chisels are shown in the drawings, with the two at the left those most commonly used. The socket chisel fits *around* the wooden handle, and the tang chisel fits *into* the wooden handle. The small mortise chisel is designed for specific mortise work, and the butt chisel for work in confined spaces.

For light chisel cuts in soft wood, confine yourself to the use of hand pressure only. For tough cuts in hardwood, use a soft-faced hammer or wooden mallet to drive the chisel through the wood.

It is mandatory that the chisel be sharp before being worked. A dull chisel is not only difficult to guide but dangerous to use. It may also split the work stock.

Generally speaking, the chisel is designed to be guided by one hand and powered by the other. When possible, hold the chisel at a slight angle to the cut, instead of straight-on. This gives a paring or sliding attack that tends to remove stock neatly and leaves the work smoother both on end-grain and ordinary cuts.

To start a chisel cut, make a defining knife mark around the section to be removed. Beginning in the center of the cutout, cut down into the stock against the grain with the bevel side to the waste, working back along the waste portion to the line. Then continue forward.

The mortise cut (Fig. 33) is the most common type to be made with a narrow chisel. Using a pencil, outline the entire mortise area on all four sides. Then score with a knife or the chisel blade along the four edges. Make a preliminary cut against the grain in the middle of the mortise to loosen the stock. Then place the chisel edge cross-grain with the bevel facing the waste wood just inside the border line. Rap the blade in. Repeat at the other end of the mortise. Starting at the first cut, and working the chisel by hand, make similar gouges about every quarter-inch across the waste area. Remove the center sections with the blade's bevel parallel to the surface.

For shallow recess cuts, make a series of close cross-grain cuts inside the knife outline and then remove the waste and smooth the surface with paring cuts.

For straight and convex cuts, use the chisel with the flat side on the work and the bevel up. Hold the chisel with the left hand and guide it with the right, applying power down on the wood and braking with the left hand. However, when cutting a long groove or dado in wide wood, turn the chisel so the bevel is down to prevent cutting too deeply.

A chisel can be used for removing rough stock from a piece of wood. In paring, or cutting curves on ends of stock, corners, and edges—both convex and concave—it is better to remove as much waste as possible with a saw before working with the chisel.

To clean the corner of a notch, hold the wood with the left hand, tilt the chisel handle away from you, and move the blade carefully toward you, using it like a knife.

To clean a concave corner, hold the bevel side of the chisel against the work, pressing down and inward.

When finished work is required, use the chisel with the bevel edge of the blade turned away

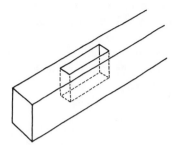

Fig. 33. A mortise is a groove or slot cut through a piece of wood, into which either another piece of wood or hardware fits. Dotted lines show mortise outline.

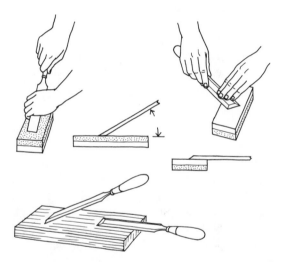

Fig. 34. To sharpen a chisel, hold it at a 30-degree angle from the whetstone on the coarse side, stroking the chisel back and forth, keeping the angle constant (at left). Turn the chisel over and remove the burr on the fine side of the stone, keeping the chisel flat down (at right). Strop the chisel on a block of soft wood as shown at bottom.

from the finished surface. The flat part of the blade will give you a good smooth surfacing.

Be careful to make the shavings thin and to cut with the grain of the wood so the surface will be left smooth and bright on any cut you make.

Never use a chisel to cut metal. Also, take particular care not to damage the blade with a foreign substance in the stock like a nail or screw. As a rule you should oil each chisel blade regularly before storing it to prevent rust.

A chisel should be used only to finish work, not to cut it. Use a coping saw for curves in thin wood, a compass saw or a keyhole saw for curves in thick wood, and a backsaw or crosscut for straight, oblique cuts.

The bevel of the average chisel is ground to a 25-degree angle. However, when sharpening a blade on a whetstone, hold it at about a 30-degree angle, pushing it along in a circular motion. Once the bevel edge is sharp, turn the blade over, lay it flat on the stone, and slide it back and forth to remove the excess metal (Fig. 34).

If a blade is badly nicked, you need not discard it. Regrind the entire edge to a 25-degree bevel, using a fine mill file or a wheel grinder. Work the mill file with a diagonal motion. Move the blade from side to side across the surface of the wheel grinder, holding it at a 25-degree angle. Cool it frequently with water. After grinding the blade, whet it as explained above.

THE PLANE

The plane is designed for two purposes: to smooth off rough surfaces and to bring woodwork cleanly down to finished size. Planes differ in size, weight, length, and width and type of blade, depending on the purpose for which they are intended. The most common types of interest to the home carpenter are the smoothing plane, the jack plane, and the block plane.

The **smoothing plane** is a most useful tool for the home handyman, since it is the smallest and lightest full-size plane. About 8 inches long with a 1¾-inch blade, it is used for finishing after a larger plane, usually a **jack plane,** has done the preliminary work.

The **jack plane** is about 14 inches in length. It is used to plane the edges of a rough board and prepare it for final truing. For the home, the jack plane is the largest plane in use.

The **block plane,** about 6 inches long and designed to be held in one hand, is used to smooth end grain, and to make bevels. Because end grain is more difficult to cut, the blade is set at an angle of 12 degrees from the horizontal, lower than

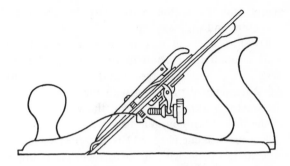

Fig. 35. A typical plane.

most planes which are set at 20 degrees from the horizontal.

Before adjusting the blade of a plane, inspect the blade by removing the lever cap and taking it out. If the blade is dull, sharpen it. A plane blade is sharpened like a chisel.

Screwed to the blade is a cap iron that should rest on the unbeveled side, slightly in back of the cutting edge. The cap iron acts as a shaving deflector during operation. The sharp edge of the cap iron and the small flat surface next to the cutter lie tight along the entire width of the blade when they are screwed together to prevent shavings from jamming in between them.

When fitting the blade in the plane, be sure to have the cap iron uppermost and on the unbeveled side of the blade. Replace the lever cap, locking it by means of the small cam at the top.

Be sure not to set the blade too far out. It is best to take off very thin even shavings no thicker on one edge than on the other. Set the blade for thin cutting, to prevent gouges in the work or clogging in the throat of the plane.

Hold the plane by the knob at the front end with the left hand bottom-side up, the bottom level with the eye. With the right hand, move the adjusting lever to the right or left until both corners of the blade project to the same distance. Then turn the adjusting nut until the blade only slightly projects beyond the bottom of the plane. You can feel the blade protruding by touching the plane's bottom lightly across the throat (Fig. 35).

Adjust the throat opening, the distance between the blade and the front edge of the opening. The throat opening determines how long a

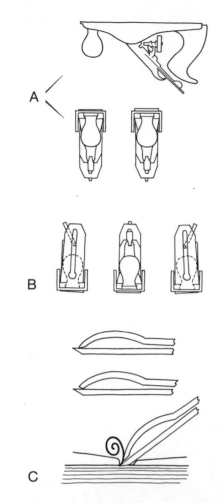

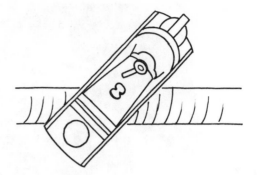

Fig. 37. Planing across end grain can be difficult. Keep the plane at a 45-degree angle to the work, and make steady movements.

Fig. 36. Hold plane as in (A) when sighting along bottom to adjust blade. Turn screw with the right hand, and hold knob with the left. Picture series (B) shows how to move the adjusting lever to the right or left until both corners of the blade project to the same height. Distance between end of plane cap and plane iron determines the amount of curled shaving to come through during planing, as shown in (C).

shaving will run without being broken; in effect, it determines how smooth the cut will be. For finer planing, the throat should be very narrow. To make this adjustment, move the frog supporting the blade forward or backward.

To plane, take a firm position in front of the work with one foot forward. Make certain that the work is held securely—either clamped in a vise with the board butting against a stop of wood tacked onto the bench, or with the board clamped to a sawhorse, bench, or table. Place the board so that you can plane it with the grain.

Hold the plane with both hands, the left hand on the knob for controlling direction, and the right on the handle for powering the plane.

Press down on the knob when you begin the stroke, then exert equal pressure on both knob and handle in the middle of the stroke. As you finish, lighten the pressure. A common fault in planing is "dubbing," or rounding the ends of the work by forgetting to lessen the pressure at the end of the stroke. Stay on top of the work as you plane.

At various stages, test the flatness of the surface in different positions with a try square. The edge of the try square should lie flat at all points. Mark high spots with chalk or pencil, and plane them away.

Work the plane diagonally from one corner to another. For large surfaces, use the widest plane available to insure a flat surface and speedier work.

For finishing, plane lightly with the grain, using a thin chip adjustment of the blade. Proceed carefully to get a smooth, square edge, testing frequently for straightness by sighting along the length of the wood, and for squareness by using a try square.

The main points in planing are to hold the tool as squarely as you can while you work, and to make sure before you begin that the cutter is sharp and set to take a fairly fine shaving.

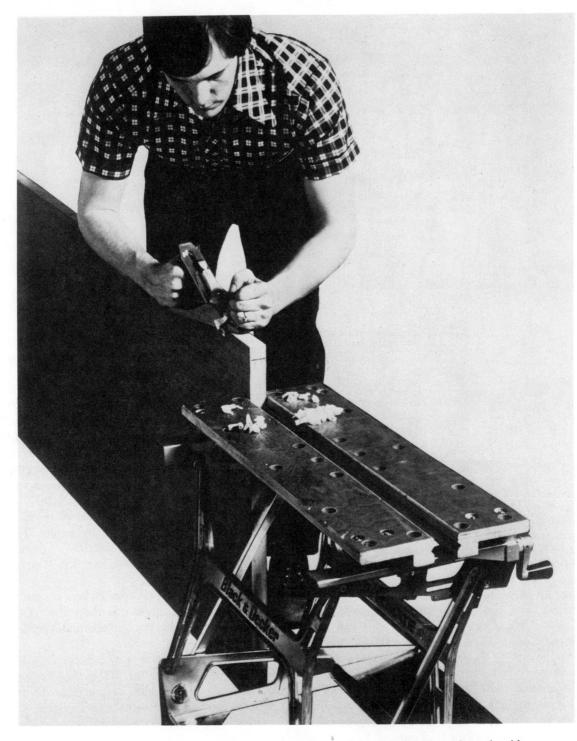

Fig. 38. For planing the side of a door, clamp the work in a vise or clamp it to the side of a workbench. Plane lightly with the grain, holding the tool as shown, and planing at a slight angle to the grain for proper cutting edge. Test the flatness of the job occasionally with a try square. Photo courtesy Black & Decker.

Planing a narrow edge presents a special problem. The thinness of the board may cause you to tilt the plane as you work and produce a slanted edge. To prevent this, clamp a strip of scrap wood in the vise alongside the work to provide a base sufficiently wide to keep the plane in position all during the planing.

When planing a larger surface, put a little more downward pressure on the knob at the beginning of the stroke than on the handle. In the middle of the stroke, both pressures should be equal. At the end of the stroke, put more pressure on the handle with practically no pressure on the knob.

The first cutting on a long, rough surface requires the use of a jack plane. Its long bottom surface, called the sole, rides over the low places and enables you to take off the high places, preserving the general plane of the surface.

If the wood has an irregular grain and is a very rough wood, plane one end of the board in one direction, and the other end in the opposite direction.

Finishing is usually done with a smoothing plane, which has a shorter sole. A junior jack plane, about 11½ inches long, will serve both purposes adequately for most household uses.

Planing an end against the grain of the wood is a more difficult task. Use a small block plane about 6 inches long. Set the cutter at a low angle so it will cut the end grain easily and will not break off the end piece.

First scrape off all foreign matter such as glue or paint with an old plane or chisel. Then fasten the work securely in a vise. Begin at the outside of the piece and plane toward the center, not making a through stroke. In the middle of the work, reverse the piece and work in from the opposite side. It may be necessary to turn the wood in the vise several times rather than approach it from the other side (Fig. 39).

In preparing an end cut for planing, saw as close as you dare to your mark, leaving only a minimum of finishing to be done with the block plane.

FILES, RASPS, AND SURFORM SHAPERS

The **wood file** is indispensible for many shaping and smoothing jobs on wood, plastic, and

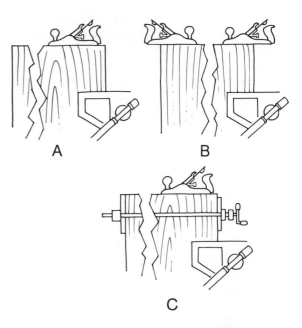

Fig. 39. Pictures show two ways to avoid splitting stock off end grain of board when planing. (A) Start planing from one side, switch direction (B) and plane in from the opposite, meeting in the middle. In (C) you can affix blocks of waste stock on both sides of wood to keep ends from breaking off.

metal. It can also be used to sharpen tools. Files vary in length, shape, and coarseness of cut. The length of the file is a matter of personal preference.

The shape of the file is designed to fit the space in which it is worked. The blade may be flat, half-round, round, triangular, or square. The coarseness of cut must be determined by the work at hand.

The teeth of most wood files are cut at an angle across the face. A file with a single row of parallel teeth is called a single-cut; one with a second row of teeth crisscrossing the first is a double-cut. The single-cut file gives a smooth finish, while the double-cut will work faster but leave a ragged surface.

Files are classified as coarse, bastard, second, and smooth—each representing a smaller distance between the lines of teeth and each producing a smoother finish to work stock.

The **wood rasp** is used on wood, leather, plastic, and soft lead. The rasp has individually shaped teeth and gives a rough cut which removes stock rapidly. It is designed for smoothing lumber, shaping small portions of wood, and for trimming joints to a snug fit. Rasps come in varying degrees of coarseness. The rasp is an essential workshop tool. Its triangular teeth cause a deeper, coarser cut than a wood file. The wood rasp is for use cutting and dressing joints in situations where the plane cannot be used, such as inner curves, recesses, and awkward corners.

The **surform shaper** is a more advanced type of wood rasp, a combination wood file and plane that can be used for shaping, trimming, and forming wood, plastics, and soft metals.

The surform comes in the shape of a file with a flat-blade surface. The blade of the surform is composed of dozens of cutting teeth like a file, alternating with holes for emptying out stock like a plane.

When working with a rasp, file, or surform, hold the tool level with your elbow, grasping the handle in the right hand against the palm, with the thumb extending on top. Hold the front end with the thumb and first two fingers of the left hand, the thumb on top. Place pressure on the tool only during the forward stroke. (One exception to this rule is the surform shaver, which cuts in the backward stroke.)

Lift the file clear on the return stroke to avoid dulling the teeth. Generally speaking, it is best to bear down only hard enough to keep the file cutting at all times. Too little pressure allows the teeth to slide over the surface, dulling the teeth. Too much pressure may chip the teeth.

Move the file in straight lines across the surface of the wood stock. Grip the work in a vise so you can hold the file with both hands. Grasp the handle in one hand and hold the point between the thumb and first fingers of the other hand.

TIPS: For finishing cuts on long, narrow work, hold the file, rasp, or surform at a right angle and move it back and forth. This is called drawfiling.

To file a curved surface, use a sweeping motion diagonally across the grain to avoid making grooves and hollows in the work.

Store the file in a hanging rack to keep it from banging against any other tool and dulling the teeth. In a drawer or toolbox, wrap each file separately in cloth or plastic.

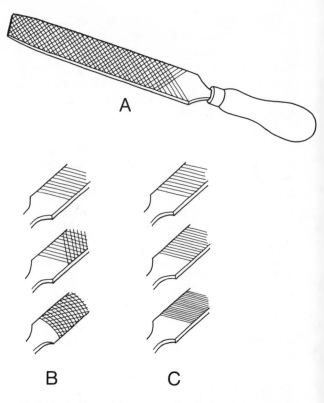

A

B   C

Fig. 40. A file and its parts (A). Column (B) shows three types of cutting surface: single cut at top, double cut in the middle, and rasp cut at bottom. Column (C) shows three types of teeth spacing: rough with large teeth at top, second cut with small teeth in middle, and dead smooth with very fine teeth at bottom.

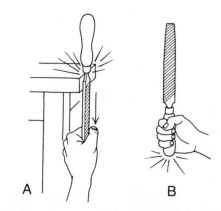

A   B

Fig. 41. To remove a file handle, hold the blade and strike it against a surface so that the handle is dislodged (A). To install a handle, bang the handle on a surface (B), forcing the blade in.

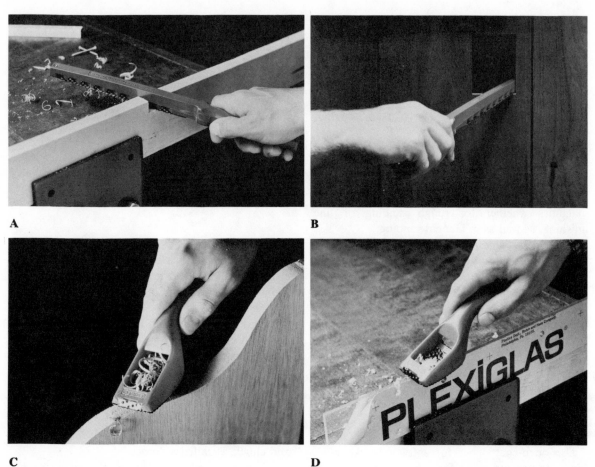

Fig. 42. One type of surform shaping tool is designed to shave, file, shape, and trim in mortises, dadoes (A), or square grooves, especially in tight square-edged areas (B). A surform will work on wood, plastics, plywood, paneling, dry wall, and even soft metal. Another type of surform is designed for cutting flat, convex, or concave surfaces (C). It will also work on plastics (D) and soft metals. The shaver works when pulled toward the operator, unlike other surforms that cut when pushed. Photos courtesy Stanley Corp.

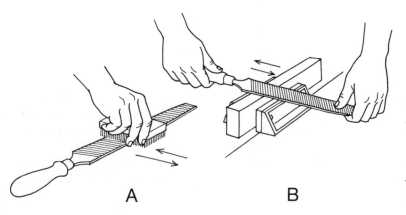

Fig. 43. Cleaning a wood file with a wire brush is shown in (A). Be sure to move the brush *with* the line of the teeth, back and forth. (B) Proper method of using a woodfile, back and forth, following the long line of the file.

SANDPAPER

Five types of sandpaper are available for the finishing of wood surfaces: flint paper, garnet paper, emery cloth, aluminum oxide paper, and silicon carbide paper. The last two are tougher, longer-lasting synthetics.

Because it is truly an all-purpose abrasive that can be used on wood, metal, or painted surfaces, aluminum oxide is the most popular of papers on the market. It can be used in all kinds of sanding machines whether for rough sanding or fine finishing, as well as in manual sanding.

All papers come in two styles: closed coat or open coat. A **closed-coat paper** has the abrasive grains packed tightly together with the entire surface of the paper covered. An **open-coat paper** has grains spread out so that there are spaces between them.

A closed-coat paper cuts faster because it has more particles per square inch. But it becomes clogged faster.

Use closed-coat paper for hardwoods like oak, steel, and plastics, where the paper can do more cutting than open coat and does not get clogged so easily.

Use open-coat paper for softwoods like pine, for removing old paint finishes, and for sanding non-ferrous metals like aluminum, copper, and so on.

For work on marble, stone, glass, slate blackboards, and ceramics, use a silicon carbide abrasive paper.

For sanding extremely rough surfaces and for removing heavy coats of paint or varnish, use very coarse grit (30) aluminum oxide paper.

To remove deep grooves and other imperfections in wood or metal and for average stock removal, use coarse grit (50) aluminum oxide paper.

For preliminary sanding on raw wood, to remove normal imperfections and small scratches, to remove light amounts of stock when creating snug-fitting joints, as well as for removing rust and other blemishes on metal, use medium grit (80) aluminum oxide paper.

For final sanding on bare wood surfaces just before the first coat of primer or sealer is applied, for removing light coatings of rust on metal surfaces, as well as for final conditioning before starting to paint, use fine grit (120) aluminum oxide paper.

For sanding between coats of paint or varnish, as well as for achieving an extra-smooth surface on hardwood which is to be given a clear or natural finish, use very fine grit (220).

BRACE AND BIT

The **brace and bit** as a tool for drilling holes in wood has almost been completely replaced by the electric drill in most home workshops. But for some purposes it is still an essential and useful tool. Holes larger than ⅜ inch in diameter are almost always made with a brace and bit. And it is still the best tool for use where there is not enough space—or no electric outlet—to operate a power drill.

Very simply, the brace and bit is a cutting edge mounted on a spindle. The spindle, complete with cutting edge, is called a "bit," and the mount, which is composed of a turning handle and chuck, is called a "brace."

When mounted in the brace, the bit is turned by the handle. A ratchet-brace is regulated by a cam ring. When the cam is turned to the right, the bit cuts into the wood; when the cam is turned to the left, the motion is reversed. For close work, especially against a wall or other obstruction, the ratchet brace is essential for drilling, allowing the operator to proceed by delicate half-turns of the handle, an impossible operation with a power drill.

A cutting bit is identified by the diameter of the hole it will cut. A set of bits from ¼ inch in diameter to 2 inches will cover most household chores.

Bits come in various lengths: the three main sizes are the short dowel, about 5 inches long; the regular, about 7 to 10 inches long; and the long ship, 10 inches long or more. The regular size is used for all-around workshop use.

Braces come in various sizes. For the home workshop, a brace with either an 8- or 10-inch

Fig. 44. Brace and bit. Dotted lines indicate the sweep of the handle, and with parts indicated. Figure at right shows the augur bit in detail.

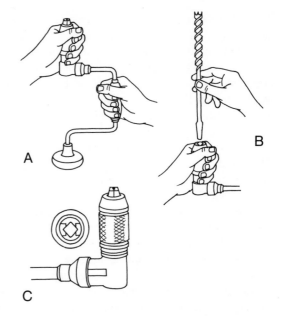

C

Fig. 46. Bit is prepared for brace (A and B). When the jaws of the chuck are open, insert the bit until it fits into the chuck (C). The inset shows how the V grooves of the bit fit into the chuck for tight hold.

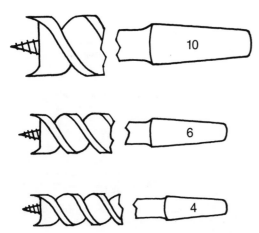

Fig. 45. Examples of bits, showing how markings on a bit indicate the size of the finished hole by gauge number. 10=⅝″; 6=⅜″; 4=¼″.

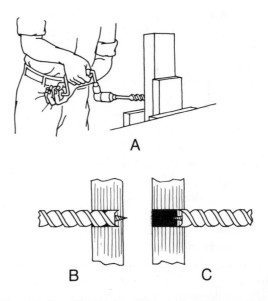

Fig. 47. Top illustration (A) shows how to clamp scrap lumber at the back of a piece of wood to keep the bit from breaking through. A second way to avoid breaking through is shown below: (B) the first step of the drilling operation, and (C) the final step from the opposite end.

sweep—the diameter of the circle described by the handle—should be big enough.

A bit is easily "chucked" into a brace for operation. Open the jaws of the chuck only far enough to admit the square end of the bit and insert it as far as it will go. When you tighten the chuck, the bit will be straight and well-centered (Fig. 46).

To bore a hole, center the bit's point on the mark, hold the round butt knob with one hand, and turn the offset handle with the other. Once the bit has almost gone through the piece of work stock, either block the piece from behind or reverse the piece and drill from the other side so as not to splinter the wood (Fig. 47).

To bore straight you should sight on the wood stock twice after the hole is fairly well started. One sight shows you whether or not you are holding the bit straight in one plane. The other sight shows you the same thing in the other plane. If you have doubts, use a square to make certain the bit is at right angles to the surface (Fig. 48).

For horizontal borings, cup the brace in the left hand, supported by the stomach. Turn the handle and give a reasonable amount of pressure on the head.

Always be sure the stock you are drilling is free of nails, screws, scraps of metal, or dirt. When not in use, cover the tip of each bit with oil and place in a toolbox out of the way.

Several accessories are available for use with a brace and bit. One of the handiest is a **screwdriver bit** which fits into the chuck and can be used to drive big screws into hardwood or into tough stock of any kind when a great deal of pressure is needed.

Screws can be driven vertically or horizontally with the bit. Blades come in varying sizes. Use the proper-size screwdriver blade for the selected screw. When working on a project where numerous screws must be driven, the screwdriver bit can be a great help in saving your hand from cramping.

A **countersink bit** is also available to fit into the brace. The flat-headed wood screw is constructed to be driven flush into lumber stock (see Fig. 67 on page 40). The countersink bit is designed to cut out a cone-shaped pocket in which the head of the wood screw sits when driven in.

For drilling very small holes, ¼ inch or less, use a breast drill, a hand drill (both of which are similar in appearance to an egg-beater), or a push drill.

The **breast drill** has a cross bar at the top to lean your chest against for pressure. The **hand drill** does not. The jaws of the hand and breast drill usually take round shank drills, but some have universal jaws that take square shanks. A set of four to six bits usually comes with a hand drill or breast drill.

When using a hand drill, do not exert too much pressure on the drill. The bits are small in diameter, and bend and break easily. If a bit is

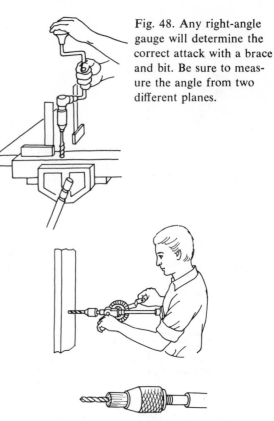

Fig. 48. Any right-angle gauge will determine the correct attack with a brace and bit. Be sure to measure the angle from two different planes.

Fig. 49. The breast drill in use, with pressure provided by chest of operator. Detail shows how to use a wood dowel with hole cut in center to govern the depth of a drilled hole. Calculate the hole's depth, measure it from the tip of the bit, and insert the dowel to stop the drill when the depth is reached.

bent, it wobbles while turning, making an oversized hole; it should be discarded.

When you have drilled the hole to the desired depth, keep revolving the drill as you remove it. If you try to pull it out abruptly, you may break the bit. Be sure to return each bit to its proper storage slot when finished, saving you the trouble of hunting down the right size next time.

The **push drill** comes in the shape of a ratchet screwdriver and is used for quickly making small holes less than ⅛ inch in diameter. A strong spring and spiral mechanism operates the chuck in a clockwise direction when pressure is applied, and counterclockwise when released. "Push" bits used with it are specially ground with points that cut when rotated in either direction. A push drill can also be used as a screwdriver and countersink with the proper bits fitted in.

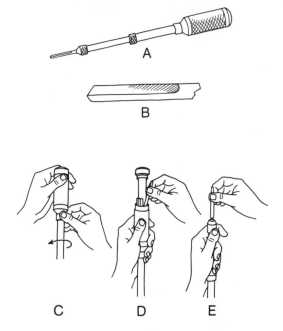

Fig. 50. The hand drill is designed for use in making small holes, like pilot holes for wood screws in plywood projects. Use a drill the size of the shank of the nail you are using; use a drill smaller than the shank of a wood screw or the screw will not hold. Keep revolving the handle when the hole is finished and as you draw out the bit. Photo courtesy American Plywood Assoc.

Fig. 51. Push drill operates with pressure exerted on handle. In (A) the push drill is shown, with (B) picturing the cutting blade of the bit. Other pictures show how to prepare drill for operation. In (C) the ring is loosened, freeing the handle, where bits are kept. Remove the proper size (D), and push down sleeve against handle (E), inserting bit into chuck.

CHAPTER THREE

Joining Tools and Fasteners

THE HAMMER

The **claw hammer** is the most common tool in anybody's toolbox. However, there are many different kinds of claw hammers, and each is designed for a different kind of work.

The two main types of claw hammers are the **curved claw** and the **ripping claw.** The curved or rounded claw is designed for removing nails and the fairly straight ripping claw is designed to pull or rip pieces of fastened material apart.

Hammer faces differ. Some are flat and others are slightly convex. The flat head is used for rough framing work where large nails are needed, and the convex type is used for driving nails flush without marring the surface of finished wood.

Handles vary in size, too. The longer handle provides more leverage in the swing than the short handle, and is usually used for framing work, whereas the short handle is used for finish work, where the blow must be more delicately controlled.

Wooden handles as well as steel handles are available for hammers. Steel handles are stronger, but the less-expensive wood handle is tough enough for use in almost any job. A wooden handle actually feels different in the hand. Choose the one you like the best.

The weight of the hammer head varies from 5 to 20 ounces. Select a weight you like but one that is not too light. A 14- to 16-ounce head is proper for finishing work, with a 16- to 20-ounce head better for framing work.

To hammer properly, grip the handle near its end. Swing it with a full stroke. Hit the nail square in the center of the hammer face.

To start a nail, hold it between thumb and

Fig. 52. The claw hammer is the most common type seen around the house. It is the main tool used by the amateur carpenter in driving and removing nails and fasteners.

forefinger near the hammer head and give it a few gentle taps with the hammer. Try to drive the nail in at a slight angle; it will hold better. Once you've started the nail, pound it in with a few hard strokes. Use a nailset to drive a finishing nail below the surface of the wood. Then fill the hole with plastic wood or surfacing putty and sand when dry (Fig. 53).

To pull out a nail, wedge the V of the claw around the nail's shank between head and wood, and pry backward on the hammer handle. To prevent marring a fine wood surface, place a putty knife or flat stick between the hammerhead and the surface. For added leverage, place a thicker piece of wood under the hammer and pry sideways, curling the nail over the side of the claw instead of straight back.

Finishing nails are usually lodged below the surface, and it is frequently impossible to remove them without ruining the wood surface. When such a piece of paneling or wood is to be removed, drive the sunken nails in with a nailset

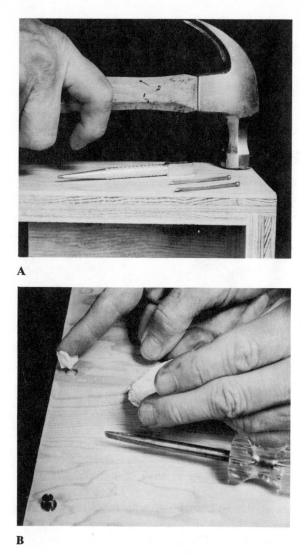

A

B

Fig. 53. (A) Shows finishing nail pounded in to bind a wood joint. Once nail is in flush, drive the head below the surface with a nailset (pictured). Always countersink screws and nails and fill the holes with plastic wood or putty (B). Apply the filler so it is slightly above the surface of the plywood, then sand it level when dry. Photos courtesy American Plywood Assoc.

past the panel or plywood thickness until the panel is free of the nail.

In fastening the vertical member of a partition to its horizontal base it is sometimes necessary to drive a nail in at a diagonal from the side through the vertical into the horizontal. This is called "toenailing." Drive the nail at an angle of 45 degrees to the horizontal and make sure that it penetrates both pieces of wood equally.

In fastening two parallel pieces of lumber together you can assure a permanent bond by driving nails through the two pieces and then bending the pointed ends flat into the bottom piece. This is called "clinching." Hammer at least three nails through the boards until the heads are firmly seated in the top board and at least one inch of shaft sticks through the bottom board. Bend the nail points in the direction of the wood grain for the best results.

If a nail bends during hammering, strike it on the edge and straighten it. If it bends a second time, remove the nail with the claw. If the second nail bends, there may be a flaw in the wood; try another spot instead. Fill the old hole with plastic wood if the surface shows.

If you are going to drive a nail into a very thin piece of wood like a molding, it is sometimes a help first to drive the nail all the way through the molding on the workbench. Then remove the nail and use the hole as a pilot hole for the final hammering in place on the wall.

A **heavy-duty nailer** can simplify complicated nailing jobs involving hundreds of nails. However, since the tool is expensive, it is advisable to rent one at a tool-supply house.

THE STAPLER

When it is necessary to do a great deal of fastening that does not require heavy nailing—like attaching insulation batts to ceiling joists—you can simplify the job by using a lightweight power-driven stapler or a manual staple gun.

You drive in each staple with a squeeze of the pistol grip. Only one hand is necessary for operation. Your other hand is free to hold the paper, insulation, or ceiling tiles being fastened.

The stapler comes in handy when you are attaching metal screening after removing rusted-out material from a door or window. The stapler can also be used for attaching certain kinds of lightweight fabrics and plastics to built-in seats or benches.

THE SCREWDRIVER

Next to the hammer, the screwdriver is the most common tool of the home handyman. Its usefulness in carpentry is much underrated.

Whenever possible, the screwdriver selected for a job should have a blade that just fits the slot in the screwhead, both in thickness and width. If the blade is too wide, it will project beyond the screw slot on either side and may gouge the surface of the work when the screwhead is flush. If the blade is too narrow, you will lose some leverage, and the tip will tend to slip out of the slot. A narrow blade will also wear the screw slot at the corners, making it impossible to turn. A narrow blade may twist or bend under heavy pressure, as well.

If the screwdriver blade is too thin, it will fit loosely inside the screw slot, will be difficult to turn, and will tend to slide sideways out of the slot.

A properly shaped screwdriver blade is ground so the tip is straight and square. The flat sides of the blade gradually taper out to meet the shank, but the taper should be so gradual that the sides are nearly parallel to each other at the very end.

Screwdrivers are sized according to the total length of blade and shaft combined. Several sizes from 3 to 6 inches in length will handle all common jobs around the house. For heavy-duty work, screwdrivers with special square shanks are also available. These allow a much greater turning force, and permit the user to apply a wrench to the shank when additional leverage is required.

To prevent screwdriver blades from slipping off screwheads, you can use **Phillips Screws** with Phillips drivers. The head of a Phillips Screw has a four-way slot into which the screwdriver fits, preventing damage to the slots or the work surrounding the screw.

For working in cramped or tight corners where an ordinary screwdriver will not fit, use a special **stubby** screwdriver. The stubby has a short blade and squat handle, and is made in both light- and heavy-duty models.

An **offset** screwdriver is also available for jobs in cramped spaces. The blade is end-bent at right angles. The offset enables you to reach into tight

Fig. 54. Ordinary household screwdriver is composed of handle, shank, and blade, as illustrated. Sizes vary in length, width, and thickness of blade, as well as types of blade.

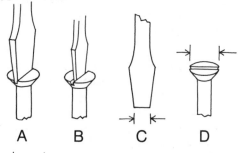

A   B   C   D

E

Fig. 55. Illustrations (A) through (D) show how to judge the proper blade for a specific wood screw. The width of the tip should equal the length of the screw slot (A). With the blade too narrow and too thin (B), the improper fit will damage the screwdriver and the screw slot. (C) Shows how width of tip should fit length of screw slot in (D). (E) Demonstrates how to use screwdriver, one hand turning the handle, the other holding the screwdriver in the slot.

Fig. 56. Phillips-type screwdriver has crossed blade for special cross-slotted Phillips Screws.

Fig. 57. For screws placed in inaccessible sites, the offset screwdriver can exert a great deal of pressure to remove or drive.

corners, while at the same time permitting application of great leverage on the screw.

If you are doing a lot of work, a spiral-type **ratchet-drive** screwdriver can save you time and effort. Pressure on this tool turns the blade in either direction, depending on how you set the ratchet control beforehand. The blade can also be locked rigidly in position for use as a standard screwdriver.

The **screwdriver bit** that you can chuck into an ordinary carpenter's brace is a great saver of sweat and energy. The bit, which comes in various sizes with either standard or cross-slot blades, is very handy for jobs where a great many large screws must be driven deeply into framing timber or hardwood.

To start a screw in cramped quarters or with only one hand available, get a special **screw-holding driver.** Little metal claws hold the screwhead against the tip of the blade while the screw is started. When the screw is in place, the claws can easily be released (Fig. 58).

You can use a grinder or file to reshape a worn, bent, or chipped screwdriver tip. Be sure to retain the original taper and keep the sides of the blade parallel near the end.

WRENCHES AND PLIERS

For the sometime carpenter, wrenches and pliers become troubleshooting tools that are turned to when special problems develop in a project. For example, when you can no longer apply enough pressure to turn a nut onto a bolt, a wrench will give you the extra power you need.

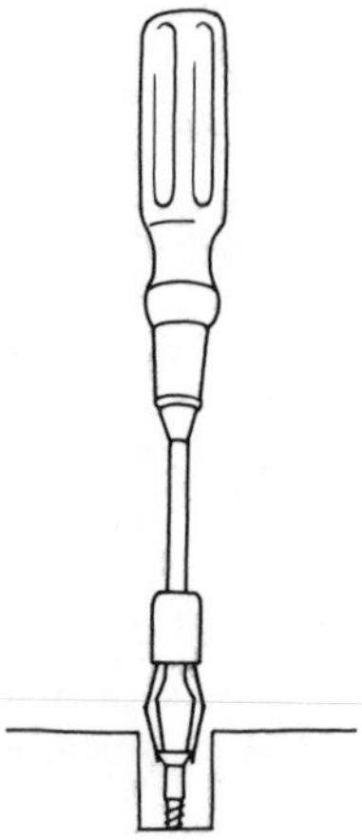

Wrenches are available individually or in sets, in both open-end and box-end styles, or in combinations of both. An **open-end wrench** is made to fit precisely one particular size of hexagonal nut. The diameter of the nut head is clearly marked on each open-end wrench. For turning a nut where access limits you from spinning it on, the open wrench is the perfect tool. A **box-end wrench** allows you to apply pressure without danger of the wrench slipping off.

If a set of open-end wrenches is not available, you can get an **adjustable crescent wrench,** which has movable jaws to fit a variety of nut sizes. When using the crescent wrench, periodically check to be sure the jaws are tight around the nut. Apply the crescent wrench correctly to the nut, with the outer crescent jaw to the right while tightening, and to the left while loosening.

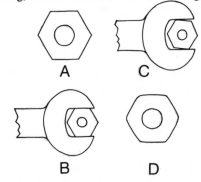

Fig. 59. The wrong-sized open-end wrench can ruin a hexagonal nut, as shown in these illustrations. (A) A good nut with sharp corners. The correct open-end wrench is secure over the nut (B). The wrong-sized wrench is applied to a too-small nut (C), with the resulting rounded corners (D).

Fig. 58. Screwdriver with specially designed screw holder grips screw tightly at point of blade, as shown in drawing. When screw has been driven into the material, a twist of the blade releases the grippers. When not in use, the screw holder slides back up over the blade onto the sleeve of the handle.

Fig. 60. Adjustable crescent wrench must always be turned in the proper direction, or injury to movable jaw will result. Picture shows right way, with movement *away* from permanent jaw.

Where a wrench can be applied from the end of the work, as in the removal of a nut, use a **box** or **socket** wrench. It operates in much tighter quarters than the open-end wrench.

For round objects like pipes, the **Stillson, pipe wrench,** or **monkey wrench,** is used. Its movable upper jaw tightens automatically when you apply pressure to the handle. You should never use monkey wrenches or pipe wrenches for turning nuts. Their jaws are designed for gripping pipes and are not parallel. The ridges on their jaws will tear nut and bolt heads to pieces.

A pair of **pliers** is one of the handiest tools around the house. However, you should not use a pair of pliers in place of a crescent wrench or an open-end wrench; you will only rip the nut or bolt heads with the teeth. Yet, in spite of a handful of no-nos, a pair of pliers is actually an "all-purpose" tool, with the applications too numerous to mention here.

There are several types of pliers for the handyman, the most common of which is the **slip-joint pliers.** The slip-joint refers to a two-position pivot that provides a normal and a wide-jaw opening. The slip-joint pliers can hold a bolt's head while turning the nut with a wrench. You can also twist and cut wire, bend metal plates, grip hard-to-hold objects, and remove broken nails and screws from wood surfaces.

Other designs of pliers include **lineman's pliers,** with side cutters for the splicing of heavy-duty wire; **channel-type pliers,** with multi-position pivots that adjust for jaw openings up to 2 inches and any shape; **long-nosed pliers,** which are used to shape wire and thin metal and often to cut; **diagonal-cutting pliers,** with no gripping jaws, which are used for cutting only; and **end cutting nippers,** which can snip wire, small nails, and brads.

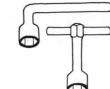

Fig. 61. The nut opening in a socket wrench can be 6, 8, or 12 points. Six-point sockets are still used for extra-heavy-duty sockets, as shown.

Fig. 62. Adjustable slip-joint pliers has double position for tight jaw as shown, or wide jaw with head slipped into other slot.

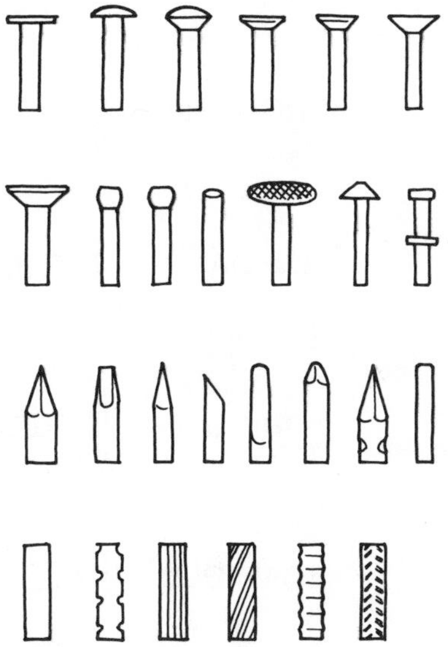

Fig. 63. Nails have three physical characteristics that set them apart from one another: the head, the point, and the shank. Here are some of the more common types of the three.

NAILS

There are at least a thousand different kinds of nails on the market today. Many of them are used for particular purposes and are beyond the needs of the average homeowner.

In household carpentry jobs, the two most-used nails are the common nail and the finishing nail, although the amateur carpenter should also be acquainted with the box nail, the casing nail, the roofing nail, the shingle nail, the screw nail, the gypsum-board nail, and others. All of these types come in varying lengths and in different gauges or body diameters.

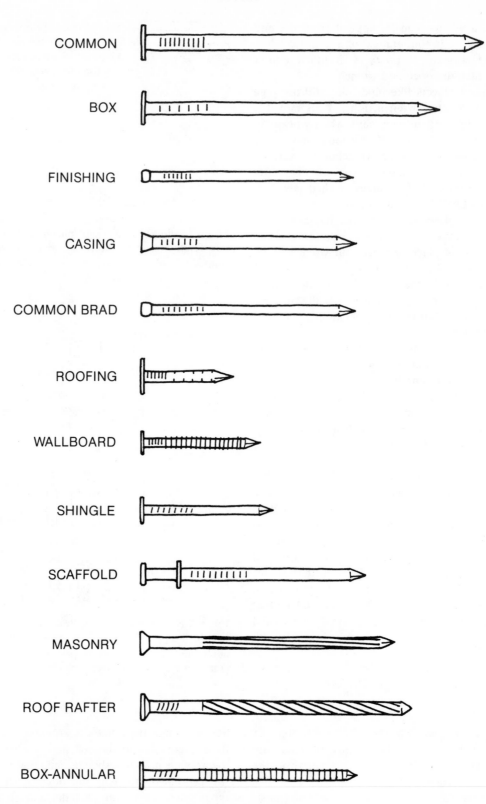

Fig. 64. Nail Chart.

NAIL SIZES

The size of a nail is measured in the old English "penny" system, based on the cost of a hundred nails of a particular size. A 2-penny nail—written 2d in the British fashion—is 1 inch long, with each "penny" size adding on another ½ inch. A 6d nail is 2 inches long; a 10d, 3 inches; and a 60d, 6 inches.

A 60d nail and larger is called a "spike," and, incidentally, will rarely be met with in ordinary household situations.

NAIL TYPES

The **common nail** is used for general construction work, and in particular for rough framing. It has a large flat head and a diamond-shaped point. The shank, or body, has grooves under the head which help increase its holding power.

The common nail is made in sizes ranging from 6d to 20d, with 6d and 8d used for subflooring, wall sheathing, and roof sheathing. Sizes 8d and 10d are used generally for toenailing studs and other 2 by 4 members. The 16d common nails are used for general framing or for attaching 2 by 4s, 2 by 6s, and 2 by 8s together.

The **box nail** is similar in appearance to the common nail, except that it is thinner in gauge and not quite as strong. The box nail is used for light construction projects—like building a box!—where a heavy nail might split the wood. The box nail comes in various sizes with either a smooth or barbed shank.

The box nail, like the common nail, is made in sizes from 6d to 20d, with 6d and 8d box nails being used for subflooring, wall sheathing, and roof sheathing; 8d and 10d for toenailing and other uses in framing; and 16d in general framing or attaching 2 by 4s, 2 by 6s, 2 by 8s, and so on.

The **finishing nail** is designed for use in a project where the appearance of a nailhead would prove out of place. It has a small rounded head that can be driven below the surface of the wood with a nailset. The hole that remains above the head should then be filled in with plastic wood or putty. The finishing nail is used in cabinetwork, in shelving, and in fastening molding and trim.

The finishing nail is made in sizes ranging from 3d to 16d. Sizes 4d, 6d, and 8d are the most common for exterior and interior trim. The same sizes can be used to install siding and paneling.

The **casing nail** resembles the finishing nail, and is used on certain types of wood trim. Thicker than the finishing nail, the casing nail has a different design in the head, which gives it more holding power.

Usually made in standard sizes from 4d to 16d, the casing nail is used for interior and exterior trim and for the installation of siding and paneling where holding power is needed as well as virtual invisibility.

The **brad** is actually a cousin to the finishing nail, with almost the same design, but is used for fine work where little or no stress is involved. The brad is made in standard sizes from 2d through 10d, and even larger for special work.

The **roofing nail** is designed with a large head and a thick, ringed shank for permanent holding of roofing paper, asphalt shingles, and wooden shakes and shingles. The head is galvanized to resist rusting. The heavy shank has a series of rings around it to keep the nail imbedded in the sheathing.

Made in lengths from 1 to 2 inches, the roofing nail has varying head diameters, in accordance with the type of installation required. It cannot be used in thin lumber, but must be attached to thick sheathing or plywood.

The **gypsum wallboard nail** is designed with a fairly small head and a long strong shank for penetration through gypsum board into framing timber. The shank usually has closely spaced rings that grip the studs to hold heavy material in place. The head is flat and fairly small so it can be pounded just below the level of the gypsum board and then spackled over. Using the head of the hammer to drive the nailhead below the surface of the gypsum board is called "dimpling."

The wallboard nail is made in sizes from 4d to 6d, the size selected determined by the thickness of the gypsum board being installed. For example, for ½-inch dry wall, use 4d dry-wall nails; for 1 inch, 6d; and so on.

The **shingle nail** comes in special lengths and head sizes designed for use with wooden shakes and shingles, as well as for asphalt, asbestos, and other types of non-wood shingles.

Most of the varying kinds of shingle nails are made in sizes from 3d to 5d. Some have thicker shanks than others, and some have larger heads. Asbestos shingle nails, for example, usually have barbed shanks and large flat heads.

The **scaffold nail** is a handy nail to have when you build a temporary structure like a scaffolding or concrete form. The nail has two heads. Drive it down to the inside head, and remove it by pulling it out by the outside head. Called also the **duplex head nail,** it is sometimes used in building stage and motion picture sets.

Made in sizes from 6d to 20d, the correct gauge is determined by the thickness of the wood used and the purpose to which the structure will be put.

The **masonry nail** is a special fastener with a variety of shank types. It is used to fasten objects to concrete, concrete blocks, soft stone, brick, grout, plaster, and other masonry surfaces. You can drive this carbon-steel nail into ordinary masonry with a hammer.

Masonry nails come with both flat and oval heads. The sizes vary from 1 inch to 3 inches in overall length.

The **screw-thread nail** is not actually a design of nail itself, but is a type of shank that can be adapted to almost any kind of nail. It can be used on a nail like the **roof-rafter nail** shown here. The screw thread tends to rotate the nail as it is driven in, and the wood fibers wrap themselves around the thread, resisting any upward pull.

The screw thread is used in many different types of nails for a variety of purposes: to fasten asphalt shingles, roofing paper, dry wall, flooring, framing, siding, wood shingles, asbestos shingles, and interior hardboard.

The **annular-thread nail** is not a design of nail either, but a type of shank that can be used in most nails. The annular thread has a series of rings or grooves close together circling the shank, like the annular box nail and the gypsum drywall nail.

Driving the nail in place forces wood fibers into the shank grooves, establishing a tight grip that resists any upward pull. The annular thread is used in the installation of subflooring and for putting together wooden frames, among other things.

A **rustproof nail** can be any type of nail made of either aluminum or stainless steel. The finish on the surface of a nail is designed to prevent rusting, but any galvanized nail surface can be injured at the head by the battering shock of a hammer blow. The result is a rusty spot that causes stains.

Aluminum nails and stainless-steel nails are expensive, but they will save you the trouble of replacement later on, particularly in wood siding or in screens that hang outside the house and cause unsightly spots in plain view of the observer.

HOW TO USE NAILS

The big danger in any kind of nailing is splitting the stock. Hardwood especially resists nailing, and even some softwood like Douglas fir, white cedar, and eastern hemlock tend to split especially easily.

To reduce the danger of splitting, use a blunt-pointed nail, or bore a pilot hole with an awl or drill, or dip the nail in wax, grease, or heavy oil.

Always select the largest nail feasible for any job, but avoid a nail so large that it creates a splitting problem. You can minimize the danger of splitting by blunting the point of the nail first by hammering on it while holding it upside down, or by filing it lightly.

When nailing, stagger heads instead of positioning them in a straight line in a job. Drive nails in so they angle toward one another, rather than all go in straight and parallel.

If a brad or tack is too small for your fingers to grip when hammering, push the brad through a piece of cardboard or heavy paper, and hold it in the cardboard until you get it started. A convenient brad holder is available at most hardware stores.

Toenailing refers to driving a nail into a stud and a base at a 45-degree angle to the base, going through the stud first and then into the sill. By nailing in a second nail from the opposite side of the stud, the toenailing is doubly effective as an anchorage measure for the stud (Fig. 65).

One of the most underrated nails of all is the **wiggle nail,** available in depths of ¼, ½, or 1

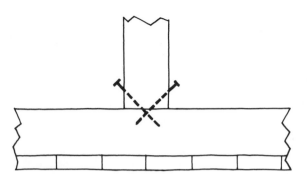

Fig. 65. Technique of driving nails into two structural members at right angles to one another is called toenailing. Nails are driven at a 45-degree angle to both stud and sill.

inch (Fig. 66). It is designed to hold together borders and frames that join at a 45-degree angle, and even to hold two pieces of wood endwise to one another.

WOOD SCREWS

The screw is a more lasting and stronger fastener than the nail, and can help a great deal if at some later date the work must be disassembled for repair, cleaning, or packing.

The common **wood screw** can be made of unhardened steel, stainless steel, aluminum, or brass; it may be bright-finished or blued, or zinc-, cadmium-, or chrome-plated. The strongest screws are stainless steel.

The wood screw has three parts: the head, slotted so that the screw can be driven into wood; an unthreaded body or shank section just below the head; and a threaded portion tapering to a point at the tip.

TYPES OF SCREWS

Classified by head style, the three most popular types of screw are the flat head, the round head, and the oval head. There are also very large screws with hexagonal heads that are driven with

Fig. 66. This metal guide is designed for convenience in driving corrugated fasteners—also called wiggle nails—in mitered joints. By holding the guide over the miter joint and inserting the wiggle nail, you can get a straight perfect set for the corrugated fastener. Photo courtesy Stanley Corp.

a wrench instead of a screwdriver. These are heavy-duty construction screws, and really act more like bolts than screws—hence their unusual design.

Most screws have a single slot in the head, but many have a cross-slotted head designed for use with a Phillips screwdriver. The Phillips Screw is used for a truly tight fastening, as in car interiors or in machinery subject to a great deal of vibration, and allows the operator to exert heavy pressure on the blade without danger of the blade slipping out.

The **flat-head screw** is installed with the head flush to the wood surface. Before the screw is threaded in, countersink the pilot hole, forming a cone-shaped space for the head to rest snug in.

The **oval-head screw** is also designed for countersinking.

The **round-head screw** is not countersunk since it is designed for installation with the head protruding from the surface for maximum holding power.

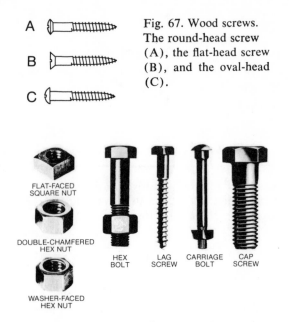

Fig. 67. Wood screws. The round-head screw (A), the flat-head screw (B), and the oval-head (C).

Fig. 68. Nuts and bolts come in many shapes and sizes, each for a specific purpose. Illustrated are a few types of common bolts, with square and hexagonal nuts shown. Photo courtesy Bethlehem Steel.

SPECIFICATIONS OF SCREWS

The average wood screw is threaded from the end point about two-thirds the length of the shank. However, the size of the screw is measured from the point to the top of the flat-head screw, and to the underside of the top of the round-head and oval-head screw.

The wood screw varies in length from ¼ inch to 6 inches. Screws up to 1 inch in length graduate in size by eights of an inch; screws 1 to 3 inches in length graduate in size by quarters of an inch; and screws 3 to 6 inches in length increase in size by half inches.

The wood screw also varies in diameter as well as length. Each screw is made in a number of shaft sizes specified by an arbitrary gauge number that represents no special measurement but indicates relative differences in diameter: 0 gauge equals $\frac{1}{16}$ inch; 20 guage equals $^{29}\!/_{64}$ inch. The gauge number is important inasmuch as it indicates the wire gauge of the body of the screw. The number can be used to determine the drill or bit size for the body hole, and the drill or bit size for the pilot hole.

The complete specification for a specific screw includes type, by head; material, or metal used; finish; length, in inches; and screw gauge, or diameter.

HOW TO FASTEN A WOOD SCREW

In all but the very softest wood, drill a pilot hole for any wood screw. In a hardwood, like oak, the hole should be the diameter of the screw's core—that is, the threaded part measured across the thread valley. The pilot hole should be bored three quarters of the depth of the screw length.

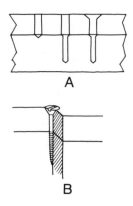

Fig. 69. (A) Shows proper pilot holes to drill when fastening two pieces of wood together with a wood screw. (B) Shows screw countersunk in place in cutaway illustration.

In a softwood, like pine, the pilot hole should be about two-thirds the core diameter and smaller still in the wood's end grain. The pilot hole in softwood should be bored as deep as half the length of the threaded part of the screw.

If the screw is soft metal, like brass, bore a

pilot hole three quarters of the depth of the screw. The hole for a small screw is usually made with an automatic push drill or hand drill.

When using long screws, drill shank-sized holes in the wood to the depth of the screw's shank—that is, the unthreaded portion. Or use a screw-pilot bit that cuts both holes at once.

On many installations a wood screw goes through one piece of wood and fastens into another. To start the screw, make a pilot hole with an awl or drill. Select a drill bit with the same diameter as the groove between the threads. For a hardwood, use a slightly larger bit; for a softwood use a slightly smaller.

The pilot hole in the second piece should be smaller—about equal to the diameter of the solid center of the threaded portion of the screw. Drill the pilot hole in the second piece of wood deep enough so that it is about half the length of the threaded portion of the screw. (A special two-stage drill that automatically bores the right-sized pilot and shank hole in one operation is available. It also cuts a recess in the surface at the same time for countersinking flat head screws.)

The proper length screw to use in a job depends on the thickness of the pieces of material being joined. The unthreaded shank or body of the screw should be just long enough to reach through the top board. The threaded portion can be as long as necessary, but it should always be at least ⅛ inch less than the thickness of the bottom board.

One of the main problems in repair involves the replacement of a screw that has pulled out of a wood hole. What usually happens is that the screw has simply torn out the wood fibers holding it in. Dip the screw in glue or paint and test it out. If the hole still won't hold, insert a wooden match or toothpick to pressure the screw in. If that doesn't work, fill the entire hole with plastic wood and let it dry. Start all over again by drilling a small pilot hole.

FASTENING HINGES

To fasten hinges in place with screws, follow this step-by-step procedure:

Locate the piece of hardware on the work in correct position. Mark the borders. Recess the work to receive the hardware if necessary, using a

Fig. 70. A. Fastening drawer pulls and door handles can be done with screws and screwdriver. Hardware comes in a variety of designs. Photo courtesy American Plywood Assoc.

Fig. 70. B. It is easy to mount surface hinges, which require no mortising and add an ornamental touch to any cabinet. On tall doors, one or two added hinges between those at top and bottom help to minimize warping. Photo courtesy American Plywood Assoc.

Fig. 70. C. Overlapping doors can be hung with semiconcealed hinges, as shown. Hinges come in a variety of styles and finishes. Photo courtesy American Plywood Assoc.

chisel to remove the stock. Place hardware in position and locate screw holes. Select screws that will easily pass through the holes in the hardware. Bore pilot holes slightly smaller than the diameter of the core of the screw. Then drive the screws tightly in place, securing the hardware.

LOOSENING TIGHT SCREWS

Tight screws can be loosened slightly by soaking them with a few drops of peroxide. There are also commercial formulations available that can be squirted at the screwhead to unfreeze it.

You can also heat the head of the screw with the tip of a soldering iron. The metal will expand against the wood, making the hole larger. When the screw cools and shrinks, remove it with a screwdriver.

ATTACHING THINGS TO MASONRY

A variety of expansion shields are available for attaching objects to masonry walls. Usually made of lead, the shields vary in size and diameter, depending on the amount of weight they are intended to support. After determining the proper-sized screw, you can fit it to the right shield. The key to good masonry installation is the proper drilling of the hole to receive the lead shield. One of three tools can be used: a star drill for a small accurate hole; an electric drill for faster work; an electric hammer for a big fastener (Fig. 71).

The **star drill** is operated by hammer blows. Diameters of star drills run from $\frac{3}{16}$ to 2 inches. For a hole larger than ¾ inch, first drill a pilot hole, then a properly sized one.

An **electric drill** with a carbide tip is faster and easier to operate than a star drill.

For a hole ¾ of an inch or larger, an **electric hammer** is the required tool. These expensive machines can be rented for the job.

Once the hole for the shield is drilled, pound the lead shield into place, insert the screw through the bracket, and tighten. The shield of lead will expand around the screw threads and fill the hole for a permanent fit.

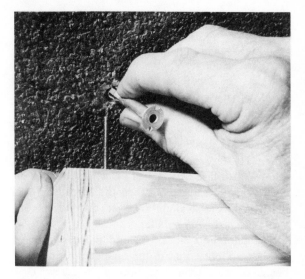

Fig. 71. One of several types of metal and lead shields designed to hold wood screws in masonry is shown being inserted in a hole drilled to the diameter of the shield's shank. Hole was made by power drill with masonry bit. Once shield is pounded into place, cabinet can be mounted by means of wood screw driven in shield. Photo courtesy American Plywood Assoc.

ATTACHING THINGS TO GYPSUM-BOARD WALL

There are three kinds of fasteners for dry-wall or plaster surfaces: angled pin, long screw, and toggle bolt.

Attaching an object like a picture or a shelf to dry wall presents a problem. It is not only sometimes difficult to locate a stud, but because of esthetic balance and space limitation, you frequently want the object hanging from a portion of the wall where there is no stud to nail or screw into.

For objects weighing 2 pounds or less, drive a straight pin downward into the dry wall at a 45-degree angle.

A conventional picture hook with a nail inserted in a canted metal holder will hang up to 20 pounds of weight.

Shelf brackets and other weighty objects should be secured by wood screws driven into wall studs through the dry wall. In places where you can't use the studs, select screws with expansion anchors or split-wing toggle bolts.

With either type, drill a hole, insert the screw or bolt in the hole, and tighten it to spread the anchor or wings that hold it from the other side of the dry-wall surface.

If the space permits, use two or three anchoring devices to secure a strip of wood to the wall, then fasten the shelf brackets to the strip of wood.

The limit of a single toggle-bolt anchor is about 100 pounds of dead weight. Anchors are not designed to secure shower grab bars or anything meant to take sudden pulls or heavy stress.

ADHESIVES

There are a number of adhesives on the market from which you can choose the proper one for each job you are doing. Adhesives vary in strength, weather resistance, setting time, and durability.

When selecting an adhesive—or glue, as adhesives are commonly called—be sure to note the use to which the work will be put. An outdoor carpentry project must be waterproof. A kitchen or bathroom project must be moisture-resistant. A basement or garage project must resist cold.

For the amateur carpenter, there are five main types of adhesive in general use, with several others that help bind foreign materials like plastics or leather to wood.

The five main types are: animal/fish glues, polyvinyl-resin adhesives, resorcinol and formaldehyde adhesives, epoxy-resin adhesives, and contact cements.

Animal/fish glue is the oldest kind of glue in use. Made of fish and animal parts and vegetable derivitives like starch and dextrine, this glue is good for bonding paper, cardboard, and leather together, hanging wallpaper, and for wood jobs where no dampness will be encountered.

Polyvinyl-resin adhesive, the familiar "white glue" that comes ready to use in a squeeze bottle, is suited to most all interior woodworking jobs and household repairs. White glue bonds strong and dries clear. It bonds paper, fabric, cardboard, cork, and leather as well as wood. It is fast-setting. When using it with wood, clamp the work with a moderate amount of pressure while it dries. White glue sets in 20 to 30 minutes, though full strength is not obtained for 24 hours.

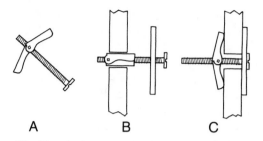

A   B   C

Fig. 72. Toggle bolt is used to secure fixtures or paintings to gypsum-board walls where stud is not in proper position. (A) Toggle bolt in open position, with the wings flared out, held there by springs. (B) Action of pushing toggle bolt through hole drilled in gypsum board or thin plywood, with fixture anchor in place on bolt. When toggle bolt has been pushed through the hole, the springs snap the wings open (C). Screwing bolt tight draws wings back to wall surface, until fixture is securely fastened.

Resorcinol and **formaldehyde adhesives** are both good glues for bonding wood to wood where structural strength is needed. Resorcinol glue is water-resistant and is used in outdoor work and boat building. Formaldehyde glue is strong, and is used for indoor furniture and indoor projects. Resorcinol adhesive is mixed with resin before application, and formaldehyde adhesive with water. Both should be clamped from 3 to 10 hours and should dry in temperatures over 70°F.

Epoxy-resin adhesive will glue almost any type of material available. Epoxy is particularly good for repairs to china and glass which are subject to use in warm water, and for strong metal-to-metal bond. It is permanent; some types set in minutes; others require 48 hours. An epoxy comes in two parts—a resin and a hardener—which must be mixed just before application. Its weather resistance is excellent.

Contact cement is used to bond wood to plastic foam, hardboard, and metal. As is obvious from the name, the cement bonds instantly on contact, with at least 50 to 75 percent of its full strength reached immediately. The adhesive is applied to the work surfaces separately, and is allowed to dry to a tacky consistency, usually in a few minutes. When the two pieces are pressed together, bond is achieved instantly. Contact cement has excellent weather resistance and sufficient strength for most home repair.

In addition to the above five types of wood-bonding adhesives are two others which are used to bond foreign materials to wood: latex-base adhesive, and mastic adhesive.

Latex-base adhesive is used to glue fabrics, carpet, paper, and cardboard, and can bond fabrics and carpeting to wood. It dries quickly and forms a strong flexible bond. It is used frequently to fasten carpeting to a wooden floor, or furniture fabric to a wood base.

Mastic adhesive is used to bond ceiling tiles, floor tiles, plywood wall paneling, or other building materials in place. **Synthetic latex** is a water-base adhesive, and **rubber resin** consists of synthetic rubbers in solvents. Both types of mastic bond to concrete, hardboard, asphalt, ceramic tiles, leather, and textiles. Mastic can be purchased in caulking cylinders and used in a gun. Weather resistance of mastic is very good.

In addition to the above, there are three other household adhesives not usually used with wood: rubber base cement, silicone sealant, and plastic cement.

Rubber base cement is used to bond rubber to rubber, paper to wallboard, and, in certain installations, wood to concrete.

Silicone sealant is used to seal the cracks around sink and bathtub where they meet the wall.

Plastic cement is used on plastic materials, and also on glass, plastic, and some types of woodwork, such as prefinished surfacing and formica-topped stock.

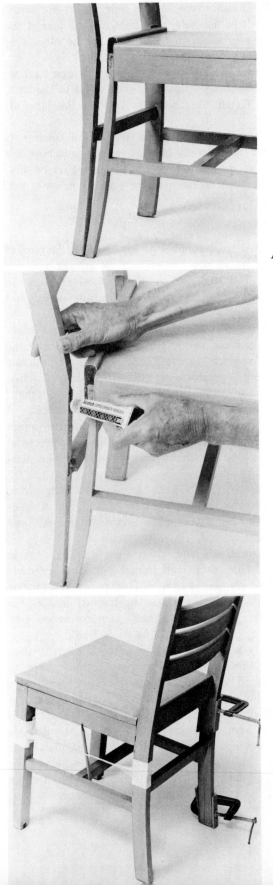

A

B

C

Fig. 73. Photograph series illustrates how easy it is to use contact adhesive on porous wood surface. (A) Shows broken chair legs. Contact adhesive is squeezed out onto both surfaces of split (B). After sides are pressed together and pulled apart again, more adhesive is applied to the two surfaces. After allowing to dry for about a half minute, parts are pressed together. (C) Two methods of clamping. Regular C-clamps are used on far leg. On near leg a splint-type clamp is fashioned out of strips of cloth tightened by twisting a pencil or strip of wood and holding it between rungs. Contact adhesive should be allowed to dry for 24 hours to achieve maximum strength on wood. Photos courtesy 3M Company.

TIPS ON GLUING WOOD

Five simple rules should be observed in obtaining neat-looking strong wood joints:

1   Pay attention to the directions on the adhesive container. Temperature conditions at the time of application are important. Be sure to clamp as necessary for the full required amount of time.

2   Scrape off all old paint and old glue before applying adhesive. Assemble only after all joints and surfaces are dry and clean.

3   Apply adhesive evenly in a smooth layer over all points of contact, using a wooden spatula or small brush. For large pieces, apply with an old hacksaw blade or a used-up windshield-wiper blade. Whenever possible, cover both surfaces with a complete coat.

4   Make all joints as snug as possible. Adhesive holds much better if the pieces it holds together fit snugly together.

5   Clamp the work firmly in position while the adhesive dries. Many different types of clamps are available (see Chapter One). Use blocks of scrap to protect the wood surfaces from damage by clamp jaws. Apply enough pressure to glue, but do not apply too much or warping will result.

On rough work, where appearance is not important, use metal fasteners, wood screws, corrugated fasteners, or even nails in addition to adhesive to fasten together pieces of stock. They need not be removed when the adhesive has dried.

On large projects, where sides are a yard apart, you can fashion a rope tourniquet to hold the ends together (see illustration of band clamp on page 11).

Epoxy is a versatile adhesive that can be used almost like plastic wood as patching or filling material for difficult repair jobs. Epoxy will fill holes in a leaking pipe. It can be used in rotted wooden gutters or rusted-out metal gutters and downspouts.

When gluing porous materials together, be sure to follow the directions carefully. Directions may advise you to spread a coating on each surface, let dry, and then spread a second coat on before joining the pieces. If you skip the second coat, the material may absorb most of the glue, leaving the joint itself without sufficient bond.

Most adhesives will not bond at temperatures below 70 degrees. Drying time is, in fact, speeded up by high temperatures for many adhesives.

In a case where two surfaces do not quite meet, mix wood sawdust with the adhesive to form a bridge between. Or use plastic wood to build up one surface before bonding.

CHAPTER FOUR

Power Tools

THE POWER DRILL

Of all the home workshop power tools, the most versatile by far is the power drill. It can be used with its many accessories for drilling, sanding, buffing, wire brushing, polishing, screwdriving, hole sawing, grinding, paint stirring, and a host of other household chores.

Power drills come in many sizes, from ⅛ inch up through 1 inch. The size refers to the diameter of the chuck, or neck, which clamps around the bit or attachment. The bit is secured by a geared-key chuck, called a Jacob's chuck, which centers and grips the bit tightly for operation.

The most common drill size is the ¼ inch, with the ⅜-inch drill running a close second. Most household jobs can be handled with the ¼-inch drill.

The ordinary drill runs on about ⅛ horsepower. For some jobs, however, a heavy-duty drill must be used. The ¼-inch drill operates on standard household current. Double-insulated models run on two-holed convenience outlets. Others require a three-hole grounded socket to prevent shock.

Some moderately priced drills have an electronic speed control in the trigger, with the tighter squeeze delivering greater speed. This regulator is a helpful accessory, particularly when you use the screwdriver bit for driving screws and need very slow turning.

The following bits and attachments are available for power drills:

The **twist-drill set** includes several small sizes, ¹⁄₁₆ through ¼ inch. The ¼-inch drill will take all sizes of twist drills up to and including ¼ inch. These drills are used to make pilot holes for

Fig. 74. Hole saw is a convenient accessory for a power drill. Here it is shown cutting through very soft metal. It can be used for large holes in plywood, wood, plastics, and soft metals. Photo courtesy Black & Decker.

screws, and other small holes. You should have a twist-drill set for everyday usage.

The **power-bit set** is available in sizes from ¼ through 1 inch, usually in a vinyl plastic holder that keeps them from knocking against one another's edges. These bits are used to make dowel holes and for cylinder-lock installation. A brace and bit can be used for holes over ¼ inch.

The **hole saw** cuts out a disc instead of merely boring a hole. The device comes with a set of blades ranging from ½ inch to 2½ inches in diameter. A pilot bit, or mandrel, holds the hole saw in place.

A **pilot bit** is made in various sizes for different wood-screw gauges. One pilot bit will drill the proper lead hole for a screw's threads, a larger hole for its shank, and countersink for its head—all in one operation.

The **countersink bit** works in wood and metal on a ¼-inch drill, providing space for the top of a wood screw to fit flush to the surface.

A **spade bit** can be used in wood and plastics with a ¼-inch drill. The bit comes in sizes from ⅜ to 1½ inches. It is used to cut fairly large holes in soft stock.

A **locater bit** is used to locate holes for screws when installing hardware. A special tube centers itself in the holes of the hardware, with the aligned bit drilling a hole in the wood.

An **extension shaft** 12 inches long is available for use in boring through thick walls, timbers, and blind spaces. A regular wood bit fits into the end of the extension shaft.

A **bellhanger's bit** is an extremely long bit—about 2 feet—used by electricians in installing doorbell wiring. It functions like an extension shaft.

A **screwdriver bit** can be used to drive screws in a power drill that is geared to turn slowly. Both regular screwdriver bits and Phillips cross-blade bits are available for quick, easy setting.

The **disc sander** is a very handy accessory to a power drill. The sander comes in two types: flexible rubber disc for curved surfaces, and a ball-jointed rigid disc for flat surfaces. All grades of sandpaper, as well as a lamb's wool or cotton polishing bonnet, can be attached to it. A disc sander performs all operations from paint removal to shaping and smoothing an irregular surface.

A **wire brush** for removing paint or rust from metal is available. It is used for cleaning tools, pans, and wood on which paint is peeling. It removes grime, rust, or tarnish, and does many jobs that would otherwise require a hand brush and lots of muscle.

A **power chisel** does the work of an ordinary chisel—cutting many familiar wood joints like the dado, mortise, tenon, and rabbet.

A **grinding wheel** can be used for sharpening tools like chisels, knives, and straight cutting edges. Grindstones or grinding wheels are abrasive stones for shaping, smoothing, or sharpening. Mount the drill on a bench, with a tool rest provided to hold the tool against the wheel. Most drills are designed to be mounted in such a stand.

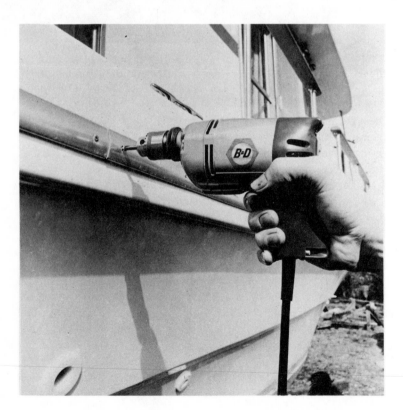

Fig. 75. Screwdriver bit can be used on a variable-speed power drill, with a regular or with a Phillips blade. Power screwdriver is most convenient when there are a lot of screws to put in or take out. Photo courtesy Black & Decker.

A

B

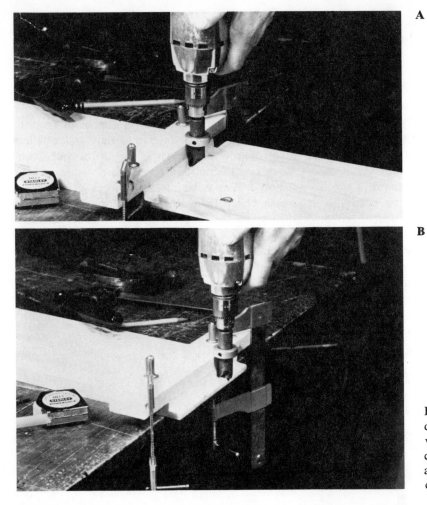

Fig. 76. Power chisel is designed to make joint cuts for various household projects. The chisel makes a dado joint (A), and a rabbet joint (B). Photo courtesy Stanley Corp.

Fig. 77. Power-bit sharpener can bring dull power bits back to life with this simplified rig. Photo courtesy Black & Decker.

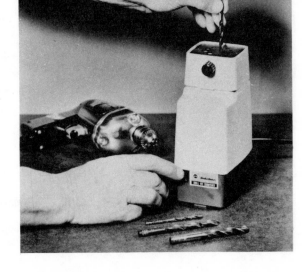

A **buffing wheel** made of layers of cloth can be used for waxing or polishing. It comes in sizes ranging from 3 to 6 inches in diameter.

A **stirrer,** which is simply a bent rod, will speed up the job of stirring and mixing paint.

A **speed control** works very well on any drill and on other small power tools. Plug the control into a convenience outlet, plug the tool's cord in it, and dial the speed desired.

USING THE POWER DRILL

The drill bit has a tendency to walk away from the mark after you start it. To prevent such movement, center-punch the drill site with an awl to provide a seat for the drill point.

Drill from a dead start. *Do not turn the motor on until the drill is in position.* If the drill is too heavy for accuracy, provide a guide by drilling the proper size hole in a waste block of wood, then clamp the drilled block into position to guide the bit. A guide is particularly useful if the hole is to be drilled at an angle.

Take precautions to prevent breakout on the underside of the hole by clamping on a backing block to the proper spot.

Align the drill bit and the axis of the drill in the direction you want the hole to go and apply pressure only along this line, with no sideways or bending pressure. Apply enough force to keep the tool cutting, but not so much that it overloads the motor.

Harder materials and larger-size holes require more pressure than softer materials and smaller holes. On ordinary workshop drills, put both hands on the drill itself for added pressure.

For drilling holes in small pieces of wood or metal, clamp them in a vise before drilling, and then apply the bit point to a previously punched mark.

On a difficult job—hardwood or metal—drill a pilot hole of smaller bore first, making the drilling of the desired size easier and more accurate.

To make drilling in metal or hardwood easier, put a few drops of oil on the drill bit. For soft metals, use paraffin.

Fig. 78. Special mount converts power drill into drill-press operation. If you're handy enough, you can make your own mount out of scrap lumber and pieces of leftover pipe. Photo courtesy Black & Decker.

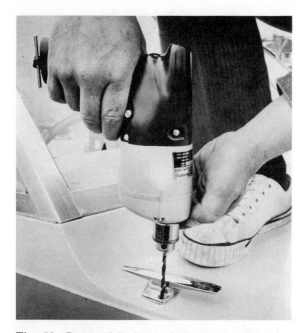

Fig. 79. Power drill is particularly adaptable and useful in drilling pilot holes for screws through hardware mounts. Pilot hole should be smaller than hole in hardware, as illustrated. Operator is using a battery-powered cordless drill. Photo courtesy Black & Decker.

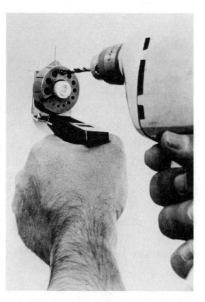

Fig. 80. One of several types of drill guides available is shown in the picture. Mark a plus sign at the spot to be drilled, align top point on vertical mark, and side wings on horizontal line. Rotate dial to correct bit size, and drill. Storage compartment holds bits in guide. Photo courtesy Stanley Corp.

OPERATIONAL TIPS

Tape the geared key for the chuck to the electric cord of your drill about a foot from the tool so you won't mislay it.

In drilling holes for wooden dowels, be sure to hold the drill at a true right angle to the surface. Place the two pieces to be fitted together accurately and mark both in position.

You can make your own doweling guide by drilling proper holes of various dowel sizes in a piece of scrap lumber.

Drill dowel holes slightly larger than the diameter of the dowels to allow space for the glue when it is applied.

Use a depth gauge to make sure the dowels will fit even when some of the space is filled with glue. If many dowels are to be used, make a cardboard gauge or jib to space them out properly. For long edgewise joinings, place dowels 6 to 12 inches apart and glue the entire edge.

THE POWER SAW

Two kinds of power saws are used in the home workshop: the circular saw and the saber saw.

The **circular saw** uses a circular blade that rotates in a counterclockwise direction as you pass it over the board. The **saber saw** has a straight blade that operates with an up-and-down motion, cutting either curves or straight lines. For straight cutting of heavy lumber, or even for ordinary boards, the circular saw is faster and more accurate than the saber saw. The saber saw is designed specifically for scrollwork, curves, and circles.

The circular saw is designed to make bevel cuts or angle cuts as well as ordinary straight up-and-down cross or rip cuts. It is a good tool for cutting 2 by 4s, large panels, and structural lumber of all kinds.

A power saw is rated according to the diameter of the cutting blade, ranging in size from a little over 5 inches to as much as 12 inches. The most popular size for the home workshop is from 6 to 8 inches in diameter. A 6-inch model will rip or crosscut a 2 by 4, even at a 45-degree angle. A 5-inch saw can be used for all kinds of trim and smaller cabinetwork. For heavy timbers, you should use an 8-inch saw, or larger.

Every power saw comes with a general-purpose combination blade which can be used either for ripping or crosscutting. However, if you need to do a great deal of ripping, you can get a special blade designed for fast ripping and longer life.

For any work with plywood panels and for a really smooth cut on regular timber that will require little or no sanding, use a hollow-ground combination blade. It cuts more slowly than other blades, but it is intended for cabinetwork, and for cutting moldings, prefinished panels, and fine hardwoods.

For cutting soft wallboard and thin plastics, use a special fine-tooth blade. For hard materials like asbestos-cement boards or plastic laminates, use a carbide-tipped blade. For sawing through masonry and metals, you can get an abrasive cut-off wheel.

A good circular saw has a spring-actuated blade guard that keeps the lower half of the saw blade covered when it is not in use. As the saw bites into the work stock, the guard is forced back out of the way. It automatically springs back over the blade when the cut is complete and the saw removed.

Certain models have a kick-proof clutch that eliminates the danger of burning out a motor. It also prevents the work from kicking back toward you if the blade binds or jams in the middle of a cut.

The portable circular saw is a right-handed tool used much like a handsaw. However, the blade cuts from the bottom up, so that the wood should always be placed with the good face down. Then any ragged edges will be left on the top, the back side of the board, rather than on the face.

With a power saw, always cut on the waste side of the line so the width of the saw cut will not affect the final measurement of the board. Be sure to support the piece being cut as rigidly as possible, preferably with two or more sawhorses or tables. Place one close enough to the line of cut to prevent vibration, slipping, or binding. If the waste piece is too heavy to support with one hand, provide a third support to keep it from splitting off good stock.

Each power saw has a depth adjustment that permits raising or lowering the base plate to control the depth of the blade's cut. Adjust this

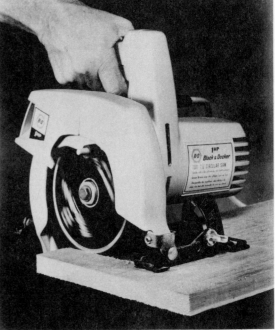

Fig. 81. Power circular saw works with teeth cutting through stock from bottom up. Most power saws can cut through 2-inch pieces without trouble. Blade can be tilted to 45 degrees, and can be set to various depths. Photo courtesy Black & Decker.

depth so that the blade just cuts through the work at the bottom, providing maximum safety combined with the smoothest cut possible.

For hardwoods and other tough-to-cut materials, set the blade deeper so that the blade clears the stock by a half inch.

The depth adjustment can be used for accurate sawing of rabbets, dadoes, and other mortise cuts involving sawing only a specific part of a piece of wood.

A variety of blades are available for use in the circular saw:

A **combination blade** is designed for general ripping and crosscutting and fulfills most of the everyday needs of the home handyman. The blade affords the fastest cutting for a variety of purposes.

A **crosscut blade** is designed for smooth, fast cutting across the grain of both hard and soft woods. This blade may also be used for rip and crosscuts on extremely hard woods.

COMBINATION

MASTER COMBINATION

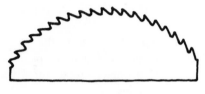

CROSS-CUT

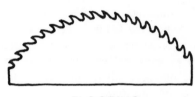

FLOORING

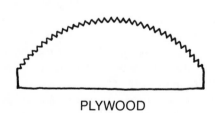

FRAMING/RIP

METAL CUTTING

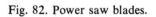

PLYWOOD

Fig. 82. Power saw blades.

A **framing/rip blade** is designed for fast cutting in all directions—rips, crosscuts, miters, and so on. This is an all-purpose blade that gives fast finishes with all kinds of woods.

A **plywood blade** has finely ground teeth to provide minimum splintering in plywood, paneling, veneers, reconstituted woods, thin plastics, and so on. Good for smooth crosscuts and miters, too.

A **master combination blade** is designed for ripping, cutting off, and mitering where a fine, smooth cut is necessary. The blade cuts so fine that sanding is unnecessary in most cases.

A **flooring blade** is designed for use with old lumber where metal objects like nails, brads, and wires may be encountered. Good for use in reclaiming lumber, opening crates, and cutting secondhand wood.

A **metal-cutting blade** is designed with teeth shaped and set specifically for cutting aluminum, lead, and other soft metals.

Not shown is a special blade for cutting steel, cast iron, stainless steel, and hard brass, aluminum, and bronze. Nor is another special blade, designed for masonry- and tile-cutting like concrete, soft brick, cinder block, asbestos, cement, concrete block, gypsum wallboard, limestone, sandstone, corrugated plastic, and soft non-ferrous metals.

The last two blades are used mostly by professionals and are not within the province of the average do-it-yourselfer.

HOW TO OPERATE A CIRCULAR SAW

Make sure the wood to be cut is securely fastened in a vise or with clamps or held under your knee and will not shift. Adjust the blade for depth of cut, then for vertical angle. On normal cuts, keep the vertical angle at 90 degrees. Be sure the blade is not touching anything. Then turn on the saw. Stand to one side of the saw when operating it, and keep your other hand well out of the line of cut. Don't hold the board with a finger out of sight underneath the board. Guide the saw along the mark by using the notch at the head showing where the blade will cut. At the end of the cut, push the saw all the way through and do not relax your grip until you turn off the saw and the blade stops revolving.

For ripping wood, measure the distance from the inner face of the board to the far side of the blade tooth and set the rip gauge. When sawing, let the gauge slide along the straight edge of the board, guiding the blade along the cut. Don't force it or it will bind.

Different woods will offer varying resistance to sawing. Oak, maple, and other hardwoods will be more difficult to saw than spruce, pine, and other softwoods. Parts of the same piece of wood may vary in density because of knots, burls, and other imperfections. Apply more pressure in denser material, letting the blade make its cut more slowly.

For a bevel cut, adjust the blade to the proper angle and tighten the nut to keep the blade in place. Go ahead as in ordinary crosscutting, following the marked guideline.

For dadoes and lapped joints, set the blade to the depth of the cut, and make a series of kerfs about ⅛ inch apart in the area to be cut. Chisel out the wood that remains standing, then smooth.

For rabbet cuts, use the same procedure as for dado cuts.

To cut narrow notches, or slots in shelf supports, make two parallel cuts to the desired depth, and then knock out the material between the cuts with a chisel. If the notch is wide, make a series of cuts as described for dadoes and proceed as above.

HOW TO OPERATE A SABER SAW

The **saber saw,** or **portable jigsaw**, is a surprisingly powerful tool that can be used to do almost everything that a stationary or table jigsaw can. Models include various sizes and styles, some of which will cut lumber up to 2 inches thick.

Usually a set of different blades comes with each saber saw—for wood, metal, plastics, rubber, leather, floor tiles, and other materials. Fine-tooth blades are used for metals and other hard materials, while coarser blades are used for ordinary lumber and thicker materials.

The saber saw can be used for starting inside pocket cuts or plunge cuts in a board without the use of a drill hole. Tilt the saw all the way forward on its nose with the front edge of the blade resting on the work, then tilt the saw backward gradually as the blade operates to sink the blade

Fig. 83. Power saber saw, also called portable jigsaw, operates a long narrow blade in up-and-down action. Teeth cut from bottom of stock upward. The saber saw is designed for making curved cuts of all types. Blade can be angled from 0 to 45 degrees from vertical. Photo courtesy Black & Decker.

into the wood. Do the job slowly. The blade will cut its way through until the saw's base is flat against the work stock, the blade projecting through the stock. The pocket cut is particularly handy when an opening must be made in the center of a large panel or when a cutout is required in a wall or a floor; for example, for sockets for electrical outlets.

Support the work firmly for use with a saber saw in order to minimize the danger of excessive vibration. Place large pieces of stock across several sawhorses, with smaller pieces gripped in a clamp or vise. When making a straight cut, clamp a straightedge along the line to act as a guide.

Fig. 84. Belt sander operates endless roll of sandpaper which continues to turn in one direction as machine passes over surface. Unit works more rapidly and efficiently then orbital sander, but is more expensive. Photo courtesy Black & Decker.

THE POWER SANDER

Power sanders fall into three main classifications: belt sanders, finishing sanders, and disc sanders. Each has its own limitations and advantages, and each is available in a wide assortment of sizes and prices.

Most powerful and fastest-working of them all is the **belt sander.** It is a machine used by the professional woodworker and cabinetmaker. Ideal for smoothing down large flat surfaces and for fast removal of heavy coats of paint or varnish, it has a continuous abrasive belt running over cylinders located at each end of the machine. On the bottom a flat metal plate holds the belt onto the work.

The belt sander is rated by the width of the abrasive belt it uses. Small sizes run from 2 to 4 inches in width, with the 3-inch belt being the most common. Special polishing belts are available for smoothing down to a final finish and for polishing plastics, metals, and similar materials.

The **finishing sander** uses a flat sheet of sandpaper clamped to a metal sole plate at the bottom of the machine. The plate has a rubber facing to provide a cushion for the paper. There are two types of finishing sanders: the orbital or oscillating sander, and the vibrator sander. The **orbital/oscillating sander** is driven in either a straight line (back and forth) or orbital motion. The **vibrator sander** has a magnetic power unit rather than a motor, and is intended only for light-duty polishing and fine finishing work.

The orbital/oscillating finishing sander is a medium-duty machine that moves the sandpaper in a flat, oval path. Working faster than the vibrator, it still manages to deliver fine finishing characteristics. The motor-driven orbital sander can handle bigger jobs than the vibrator, and is actually an all-purpose sanding machine for the homeowner.

The finishing sander has a flat rectangular pad for holding the abrasive paper, permitting you to get into corners where a belt sander will not go. The paper is held by clamps at each end. The paper for a finishing sander is cheaper than paper for a belt sander since all models are designed so that standard sheets can be cut down to the size needed.

For thin veneer woods, the orbital sander has a slower cutting action and is safer where a high-quality finish is needed. It can be used for final polishing of plastics, metals, and finished surfaces. Special polishing pads are substituted for the usual abrasive sheet on the pad during polishing sequences. The orbital sander is actually not practical for removing paint or for finishing rough lumber. For such a job, the belt sander should be used. The amateur carpenter can rent a belt sander for such a job at a tool-supply house.

The **rotary-disc sander** works from a spindle that rotates a round flat pad holding the paper or

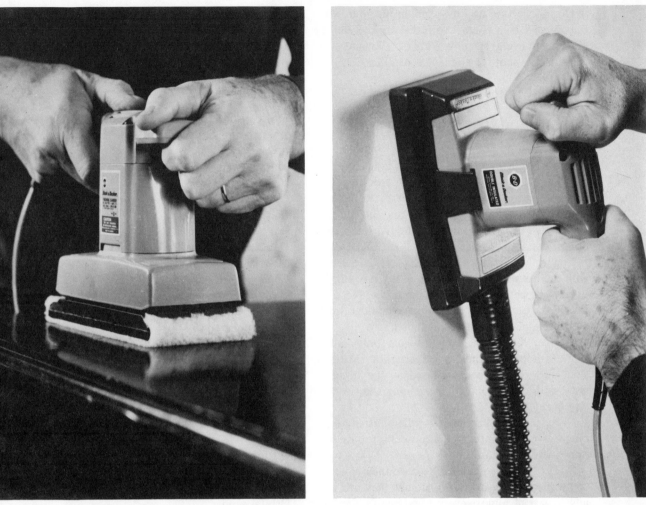

A   **B**

Fig. 85. Orbital/oscillating power sander pictured with special accessories. (A) Polishing attachment. (B) A clamp-on cover that sucks up sawdust and evacuates it through hosing piece. Orbital sander can be used to smooth finish on woodwork, paneling, flooring, and furniture. Photo courtesy Black & Decker.

polisher. The disc takes circular pieces of sandpaper varying in size from 4 to 6 inches in diameter. The abrasive disc is usually attached by a screw that fits into a recess in the center of the disc.

The disc sander usually consists of spindle-mounted accessories that can be clamped into the chuck of an ordinary electric drill (see Chapter Four).

Regardless of the type of sander used—belt, finishing, or disc—do not bear down excessively hard while working the tool. Extra pressure slows down the action of the paper and makes it cut more slowly than it should. In addition, pressure causes undue wear on the abrasive paper, as well as on the motor and bearings of the machine.

Start each sanding job with the coarsest grit, then work progressively down to the finest grit necessary. Use a sander with the grain only, wherever practical, especially on the final finish.

Do not start or stop the sanding machine when the abrasive is in contact with the work. Keep the machine moving steadily without allowing it to stand in one spot for any length of time.

THE BASIC CARPENTRY TOOLBOX

Every handyman has a different conception of the contents of a basic carpentry toolbox. His particular choices are often dictated by his own special talents and interests. Or his choices may be dictated by the type of home he lives in—large house, cottage, or apartment.

The following basic carpentry set is selected with the average homeowner in mind, meeting ordinary carpentry situations in the home. No tools involved strictly with plumbing, electrical work, or heating-plant repair are included.

Devices like nails, screws, toggle bolts, adhesives, sandpaper, and so on are not included, since they should be purchased as the occasion rises, with each problem determining the size and type needed.

Nor are paint brushes or rollers listed, inasmuch as each painting situation calls for its own implements and finishes.

Power tools *are* included because they have become so common and familiar that they are usually a part of almost every handyman's toolbox.

Of course, the work they perform can be taken care of with hand tools in the list.

Make additions to the list as you meet special problems. With this group of tools, you should be able to face most ordinary carpentry situations around the home.

Flexible steel rule, 8'   (or folding rule, 8')
Combination square   (or steel square)
Level
Marking gauge
Vise
C-clamps
Claw hammer, 16 oz.
Nail set, $\frac{2}{32}''$
Handsaw, crosscut, 26''
Miter box
Coping saw
Utility knife
Chisels, ¼'', ¾''
Smooth plane, 8''
Rasp   (or wood file)
Surform shaping tool
Brace, 10''
Bit set, ¼''–1''
Hand drill
Crescent wrench   (or open-end wrench set)
Pliers
Oilstone
Power drill, ¼''
Power saw, circular, 6''
Orbital sander

Section II

MATERIALS

The amateur carpenter must be as familiar with the materials he is using as he is with his tools. He must also know how to procure the material in such a way that he doesn't waste money or buy the wrong type for the job he is contemplating.

This section on building materials is divided into two chapters.

The first, Chapter Five, on shopping for lumber, tells how to pick out the kind of wood you want and how to order it with minimum waste of material, time, and money.

The second, Chapter Six, is a rundown on the various types of building materials now available to the homeowner for construction and for repairs, including both wood and wood substitutes developed in recent years.

CHAPTER FIVE

Shopping for Wood

Although homes can be made of stone, brick, or other types of masonry, the majority of them are built of conventional wood-frame construction inside and out. Because wood plays such an important part in the average dwelling, you should know what species are available for repair purposes, how to select the right pieces for a specific job, and how to shop for lumber.

Wood comes in a wide variety of types: hard, soft, smooth, rough, thick, thin, long, short, stiff, and flexible. There is a right kind of wood for every project.

Two ways to save money in lumber are to buy the lowest grade and the lowest-priced wood that will do the job, and to buy the smallest quantity possible. The first requires knowledge of construction techniques, and the second calls for accurate measurement of needs.

When you take on a project involving structure, like a new room, barn, garage, or shed, you must use lumber that is adequate in strength, a fact that is more important than appearance.

Grade stamps required now on all framing lumber are the key to the capability of a piece of lumber. The stamp shows the grade, the species, moisture content, and identifying number of the producing mill. It is determined by the American Lumber Standards Committee, which has an over-all standards enforcement responsibility upheld by government and the courts.

The grade stamp tells size in width and thickness at the nominal figures only. For example, a 2- by 4-inch piece actually measures 1½ by 3½ inches when thoroughly dry. And a 2 by 8 is actually 1½ by 7¼ inches. Knowledge of these

differences becomes important to you on all types of wood projects.

It is likewise essential that you understand enough about the appearance of wood to select the right type for whatever project you have in mind. Spending a fortune on a strip of mahogany for a closet shelf that might be better built of soft pine is a waste of money. On the other hand, a grade-C plywood surface hung on a living room wall would be a fiasco of economy.

Knowing how to shop for wood demands knowledge and comprehension of lumberyard terms. You must know the advantages and limitations, along with the peculiarities, of lumber. Only then can you get the best for your money and the right material for the job.

TYPES OF WOOD

There are two basic kinds of wood—softwood and hardwood. Softwood comes from needle-bearing trees like pine and fir. Hardwood comes from broad-leaved trees like oak and ash. "Hard" and "soft" are misnomers: some so-called soft woods are actually harder than hardwoods, and vice versa. Because of the nature of the wood and of the uses to which each specific species is put, softwoods and hardwoods are cut and graded differently.

For repair work in the average home, softwoods are usually used: Douglas fir, spruce, cedar, and so on. However, if the repair entails work on floors and stair treads, hardwoods like maple and oak are usually used.

Among the general points about lumber you

should know are these: seasoning, grain characteristics, defects, size, type, figuring board feet, classifications, and grading.

SEASONING OF LUMBER

At the time a tree is felled, its weight is composed of from 25 percent to 70 percent water on the basis of the total weight of the wood in "green" or uncured condition. Green wood will warp or split with subsequent drying out. Not until wood dries, loses weight, and shrinks does it increase in strength and become usable lumber.

This process of drying out is called "seasoning." Seasoning will reduce the water content of wood by 6 percent to 20 percent, depending on the intended use of the wood.

Lumber can be seasoned by air drying, by forced drying in a kiln or oven, or by a combination of the two processes. When you use unseasoned wood, you may find that it shrinks after it is in place, that it warps, and that it may not take a coat of paint.

Lumber purchased at a dealer's is pretty well seasoned, although it may still be somewhat green.

Grade stamps indicate "S-DRY" or "S-GRN," showing that the lumber was dried before being surfaced, or smoothed, at the mill (S-DRY), or that it was surfaced before drying (S-GRN). Lumber can be kiln or air dried, but to qualify as S-DRY the moisture content must be down to 19 percent or less. "Dry" lumber has already shrunk down to the size it will maintain in construction.

A mill may surface its lumber "green" or unseasoned, but the size has to be enough bigger so that when dried out naturally it will become the same size as the "DRY." Dry lumber is much stronger than unseasoned.

For most work the amateur carpenter will prefer seasoned lumber. "S-GRN" lumber in the dealer's yard is virtually dry, because of its time in storage, but you should check the time stored when you purchase it. Dry lumber can withstand a severe soaking in the rain before its dimensions change measurably. Some lumber carries the mark "MC 15," which means the wood was dried to 15 percent or less—but for most work around the house this isn't necessary.

For home repair work like closets, pantry

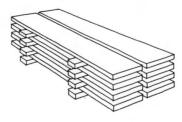

Fig. 86. Lumber can be stockpiled in many ways when it is delivered outdoors to the building site. Illustration shows one good way to protect boards from weather. Be sure lumber does not touch ground. Western Red Cedar Lumber Association.

shelves, and drawers you can use any kind of commercial lumber. And you should expect some warping or twisting. But for trim or molding buy only well-seasoned material. Shrinkage in trim will make ugly gaps between door and frame and window and frame.

HOW TO SEASON WOOD INDOORS

If you buy wood that looks pretty green, you can give it a quick-dry seasoning by stockpiling it in your house for a few weeks or a month before using it. Or put it in a dry cellar, a well-ventilated attic, or the garage.

To stockpile, lay the lumber on traverse 2 by 4s spaced not over 4 feet apart. Then place separators of lath or 1-by-2 strips between the boards to provide air circulation. A few days before you are going to install the wood, move it into the room where it will go up. You can acclimate the wood to its future home in this manner and insure that it will suffer minimally from shrinking and splitting.

CHARACTERISTICS OF LUMBER

The way wood is sawed at the mill gives lumber special characteristics. It can be cut from a log at right angles to the growth rings, or tangent to the growth rings. Right-angle cuts are called "edge-grain" or "vertical grain" in softwoods, and "quartersawed" in hardwoods. Tangent cuts are called "flat grain" in softwoods and "plainsawed" in hardwoods.

Right-angle cuts keep shrinkage or swelling to a minimum. For paneling, however, tangent cuts are preferable, particularly if you are looking for a striking pattern on the surface of the wood.

DEFECTS IN WOOD

The way lumber is seasoned can cause defects. If it is seasoned too slowly, it may develop stain spots. If it is seasoned too quickly, it may warp. Warped boards are hard to work with, and may split at the time they are nailed up or even afterward.

Besides decay spots, the most obvious defects in lumber are:

Knots. The cross-sections of branches, ranging from *pin size* (less than ½ inch) to *large* (over 1½ inches), *tight* (firmly embedded), or *loose.*

Decay. Disintegration of the wood fiber, usually caused by fungus growth helped along by excess moisture.

Dote or **doze.** The same as *decay.*

Stain. Discoloration, brown or blue.

Pitch pockets. Chunks of pitch embedded in the wood.

Wane. Bark left on the edge of a board after trimming, or a gap in the wood caused by displaced wane.

Bark pockets. Bits of bark enclosed in wood.

Skips. Rough or undressed areas caused when the plane misses a portion of the surface.

LUMBER SIZES

As for the designations of lumber sizes, the nomenclature is quite simple once you understand the system. Lumber is described in this order: by thickness, width, and length. Thickness and width are stated in inches, but length is stated in feet. A 2 by 4 is 2 inches thick and 4 inches wide at the time it is cut from the green log.

Actually, when you get a 1 by 6 finished piece, it is really ¾ inch by 5½ inches. In milling, a quarter of an inch was shaved off the flat sides, and off each end. A finished 4 by 4 is actually 3½ inches by 3½ inches.

Tongue-and-groove lumber (which is fitted on the edges, Fig. 89) requires more milling at the

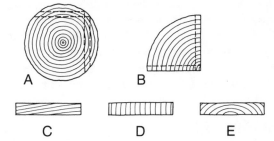

Fig. 87. Different wood grains come from different cuts in the green log. Lumber from the outer perimeter is called "plain-sawed" in hardwoods (A). In a softwood, the cut from the tangent is called "flat grain" (B). Sometimes the grain is "indefinite" (C). Lumber from the inside of the log is cut at right angles to the middle, or "quartersawed" in hardwoods (D). Boards from the inside are called "edge-grain" and "vertical grain" (E). Floor strips of good quality are usually quartersawed. Paneling with interesting patterns is usually plain-sawed. Vertical grain is more expensive than flat grain or indefinite.

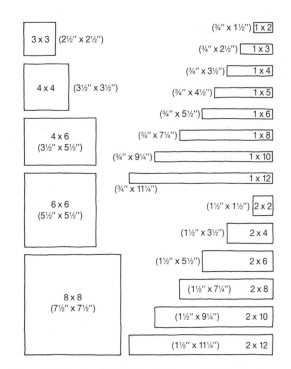

Fig. 88. Sizes of wood widths and thicknesses are misleading since a 2 by 4 is really only 1½ inches by 3½ inches and a 1 by 8 is really only ¾ inch by 7½ inches. The amateur carpenter must know the exact measurement if he is to work successfully with wood. Chart shows true dimensions of typical cuts.

edges in order to make the tight-fitting joints. Because of this, each strip is narrower than the standard size mentioned above. When you are figuring the amount of wood to order for a wall area or floor section, you must take into consideration these deviations from the order size.

BOARD FEET, RUNNING FEET

In purchasing wood you are billed by the board foot. A **board foot** of wood is a piece of rough, green, unfinished wood 1 inch thick, 1 foot wide, and 1 foot long, or an equivalent volume. In other words, a 1- by 6-inch strip of wood 2 feet long is a board foot of lumber, as is also a 1 by 4 piece of framing lumber 3 feet long.

To find out the number of board feet in a given piece of lumber, simply multiply the thickness in inches by the width in feet by the length in feet. A 2 by 4 that is 6 feet long would be multiplied out as $2 \times \frac{1}{3} \times 6$, or 4 board feet. Remember that the width must be divided by 12 to be converted to feet.

Sometimes you buy certain pieces like trim by the **running foot**, or the **linear foot**, which means exactly what it says. You buy 10 feet of 1 by 8 at so much a foot. A 10 foot length of 1 by 8 at 16 cents a running foot would cost $1.60.

In ordering wood at the lumberyard, specify the actual length of each given board. If you want 14 pieces of 1 by 12 No. 2 white pine, each 10 feet long, the shipping ticket reads: "14/10 1×12 com pine 140 feet 250M $35.00." The number of board feet is 140 (obtained by multiplying $1 \times 1 \times 10 \times 14$). The price in dollars per M (thousand board feet) is 250, from which the charge amount of $35.00 is obtained by multiplying $\frac{140}{1000} \times \$250$.

NAMES OF LUMBER SIZES

Many sizes of lumber are prepared for industrial usage, but the wood you will be working with in your home is called "yard" lumber. Terms like *board, timber,* and so on each refers to a specific size, and are defined as follows:

Strip—a piece of wood 1 inch thick but less than 6 inches wide.

Board—a piece less than 2 inches thick and 6 or more inches wide. Fencing, sheathing, sub-flooring, roofing, concrete forms, and box materials are boards.

Dimension—a piece from 2 to 5 inches thick, and 2 or more inches wide. Framing, joists, planks, rafters, studs, and small timbers are all dimension stock.

Timber—a piece 5 or more inches in the least dimension. Beams, stringers, posts, caps, sills, and girders are all timbers. In practice, a timber can be a combination of dimension lumber nailed together in parallel.

GRADING OF LUMBER

The grading of all woods according to quality is done according to rules written to cover individual characteristics of species in performance and appearance.

For permanently visible uses, quality and appearance become dominant factors in selecting lumber. Utility-grade 2 by 4s are allowed in most construction, and they're strong enough for a garden pergola, toolhouse, or arbor, for example. But you may prefer the more uniform good looks of the two higher grades—Select and B and Better, explained below—with smaller knots or none at all. You'll have to decide whether the looks are worth the extra dollars. For garden structures, carports, and the like, unseasoned lumber is satisfactory, and lower in price—if it's available locally.

Boards usually are not grade stamped, as the stamp ink leaves a mark that is difficult to remove. However, boards definitely are graded at the mill, and the dealer orders them by grade. He usually keeps different grades in separate bins.

Grading is done mostly on appearance, and you can choose your boards for looks also. But you might prefer a knotty grade, even though it might not be as strong for heavy-duty shelving as a clear grade.

Dimension timber, which is lumber used for construction and is rarely seen as a visible surface, is graded according to natural characteristics that affect its strength, stiffness, and general suitability. Usually dimension is sub-graded 1, 2, 3, and sometimes 4, representing the range of qualities for joists, studding, and bracing. The top grade carries more load, but all grades are suitable for most light-construction projects.

SOFTWOOD GRADING

Select grades of softwood are the best grades. Wood graded "select" is used for finishing purposes, and is broken down into four sub-grades usually designated by letter. The two highest sub-grades of select, A and B, are combined and sold as **B and Better Select.**

The high select grade is usually a clear, or practically clear, wood. B and Better Select (B & Btr) is stamped on the highest quality of interior and exterior finish, trim, moldings, paneling, flooring, ceiling, partition, beveled siding, and drop siding. **C and D Select** grades can be used where saving is considered more important than perfect appearance. All select grades, including C and D, take natural finishes well.

Common grades are the next step. These woods are used for utility and construction purposes. Common grades are sub-graded 1 to 5, or 1 to 3. You can use No. 2 "common" pine or No. 2 "common" cedar if you are interested in letting the knots show for informal effect in a paneled wall. A good common grade of some species of softwood will give you a satisfactory surface for paint. No. 1 is of course the best grade.

As a matter of fact, the best thing to do is to use the lowest possible grade you can for the job. Many times you can get twice as much lumber for the same number of dollars. Your local retail lumber dealer can give you the best practical advice. Just tell him what you are going to use the wood for, whether it is to be painted or not, what tools you have, and your own special preferences.

GRADES OF WOOD SIDING

Solid wood siding comes generally in three grades, with **Clear** at the top. Most widely used are the other two, **⚹1 Common** and **⚹2 Common,** or **Select Merchantable and Construction** as they are sometimes called. A tight-knot ⚹2 Common grade may prove quite acceptable in appearance, and its cost is lower.

The most widely used siding patterns are bevel, channel rustic, tongue and groove (flush edge), and plain edge boards, but with battens (Fig. 89). Bevel siding is for horizontal application;

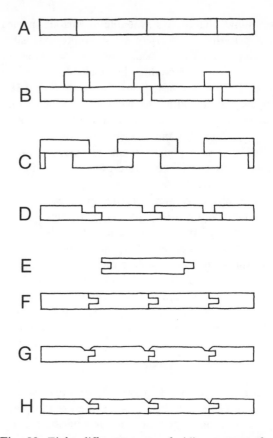

Fig. 89. Eight different types of siding patterns for solid-wood stock are shown: (A) ordinary vertical planks; (B) typical board-and-batten, with the small strips on the outside; (C) board-on-board, a variation of board-and-batten; (D) channel groove, an effect that gives a deep-shadow silhouette between strips to accentuate their interest (application here is shiplap); (E) a tongue-and-groove panel with several different designs; (F) flush tongue-and-groove; (G) channel, or eased channel, with a modified channel pattern; and (H) V-joint, sometimes used on interior paneling as well. All these patterns can be used inside the house on interior walls.

board and batten is for vertical. Tongue-and-groove usually is used vertically, and channel rustic is applied either way. Siding often has one side surfaced (smooth) and the other side rough (bevel style is the best example), which allows a choice in surface texture. Rough side is popular for staining, smooth for paints.

Board paneling for interior use is selected mostly for appearance only. The lumber retailer will show what he has available in various species and patterns. The variety can be quite extensive, ranging from dark woods, such as western red cedar, to light, including hemlock, Idaho white pine, and other pines; from narrow boards (3½ inches actual width) to wide (11¼ inches actual); in several patterns such as shiplap, flush joint, and tongue-and-groove joints with figured edges, and with planed or sawed surfaces. Some siding patterns make interesting interior panelings.

It usually takes more time to apply boards to a wall than 4- by 8-foot panels, but some people are aware of the difference and find the added effort worthwhile. Boards reflect greater handcraftsmanship, and veneer-faced panels never have duplicated the solid look of lumber. Furnishings and a family's tastes will influence the choice of board width and pattern, species, and whether the finish will be paint, stain, or natural.

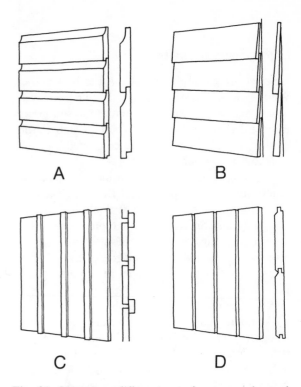

Fig. 90. How four different exterior types of wood covering look. (A) Horizontal drop siding, with a diagram of its profile. (B) Horizontal bevel siding. (C) Vertical board-and-batten application. (D) How tongue-and-groove siding in vertical application looks. Tongue-and-groove can also be applied in horizontal fashion.

PLYWOOD GRADING

Certain grades of plywood—notably N and some A (explained below)—can be used in place of conventional sheathing as an exterior surface and as an interior surface without the application of any other finish material or paint over the surface. Texture 1-11 is an example of an outdoor finish siding material that needs no paint.

Other grades of plywood are unsuitable for anything but underlayment—that is, the layer beneath a finished floor or sheathing under an exterior application of wood, brick, or stucco. Plywood is as varied as it is versatile and must be understood to be used to best advantage.

Thus, in shopping for plywood, you should know something about the various kinds of ply available, and you should know how to select the kind you want for the specific purpose you have in mind.

TYPES OF PLYWOOD

Plywood is manufactured in two basic types—exterior and interior—with a variety of appearances and quality grades within each type. Exterior differs from interior plywood only in the kind of glue bond and the grade of veneer used. The adhesive used in exterior plywood has a completely waterproof bond, whereas the adhesive used in interior ply is made of highly moisture-resistant or waterproof glue.

Plywood is made both from hardwoods and softwoods. It is the outer layer of the sheet—called the **face veneer**—that determines its class.

Softwood-plywood face veneer can be Douglas fir, hemlock, cedar, larch, spruce, redwood, knotty pine, white fir, and white pine.

Hardwood-plywood face veneer can be an endless number of woods, each of which is usually milled and selected to accentuate the natural beauty of the hardwood.

Every piece of softwood plywood is graded with a letter standardized by the American Plywood Association, an organization of plywood manufacturers. The letters are N, A, B, C, C (plugged), and D. Each piece of softwood plywood is also marked either EXT or INT (exterior or interior).

N is a special-order "natural finish" veneer, very select, made of heartwood, free of knots, splits, pitch pockets, and any open defects. A few small flaws are permitted, if well matched.

A is quality veneer with highest standard veneer appearance. There are no open defects, and the surface is suitable for light stain-glaze finishes. All patches and repairs are neat and run parallel to the grain.

B is similar to A veneer, but circular plugs are allowed. Sound tight knots up to 1 inch and splits not wider than $\frac{1}{32}$ inch are allowed.

C is the lowest quality veneer permitted in exterior type of plywood. Knots up to 1 inch, borer holes, splits $\frac{3}{16}$ inch and less, and tight knots up to 1½ inches, plugs, patches, shims, and minor sanding defects are allowed.

C (plugged) is used for underlayment. Underlayment is material used under floor finish surfaces, or under any other type of finish surfaces. Tight knots up to 1½ inches and worm and borer holes not more than ¼ and ½ inch are allowed.

D is quality veneer used only in interior-type plywood panels. Knots up to 2½ inches, pitch pockets, limited splits, worm or borer holes, minor sanding defects, and repair patches are allowed.

A panel marked *A-A* has the highest standard quality veneer on both face and back panels. *A-B* means there is A appearance on the face panel, B on the back.

A special type of exterior paneling which can serve as the finish for an informal home is called Texture 1-11 (One-Eleven). It has an exterior plywood with deep parallel grooves, a usually unsanded surface, and other rustic natural wood characteristics.

Other surfaces are shown in Fig. 92.

HARDWOOD PLYWOOD GRADINGS

Hardwood plywood differs from softwood plywood in the particular use to which it is put. Softwood plywood is used for sheathing, flooring, roofing, and so on. Hardwood plywood is used for interior wall covering, desks, cabinets, and other such fine work.

Hardwood face veneers which are sold with prefinishing already applied are now used extensively in homes. You can either nail them in place or glue them over existing walls with special adhesive. Once installed, a prefinished plywood panel needs almost no maintenance other than an occasional polishing. A wide variety of finishes is available.

Hardwood plywood is graded according to standards set up by the Hardwood Plywood Institute and stamped onto the wood, as is the type of adhesive used.

Custom grade is made especially to the specifications of the architect or builder, and has matched grain panels.

Good grade **(Grade 1)** is used where appearance is important, and where natural finish is desired. The joints are matched, and no knots, wormholes, splits, or decay are allowed. This is the top grade of consumer hardwood plywood.

Sound grade **(Grade 2)** is free of open defects. The veneer is not matched for grain or color, and is intended for use as a paint surface. It may contain knots up to ¾ inch, but has no forms of decay.

Utility grade **(Grade 3)** is a strong serviceable

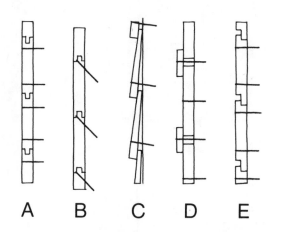

A   B   C   D   E

Fig. 91. Different nailing techniques are used on various solid-board sidings. Drawing (A) shows face nailing of tongue-and-groove. (B) Blind-nailing of tongue-and-groove. (C) Proper nailing for bevel siding, with nail angled just slightly down. (D) Typical nailing of board-and-batten, with nail at batten, and another in middle of board. (E) Face-nailing for channel-groove shiplap-joint siding.

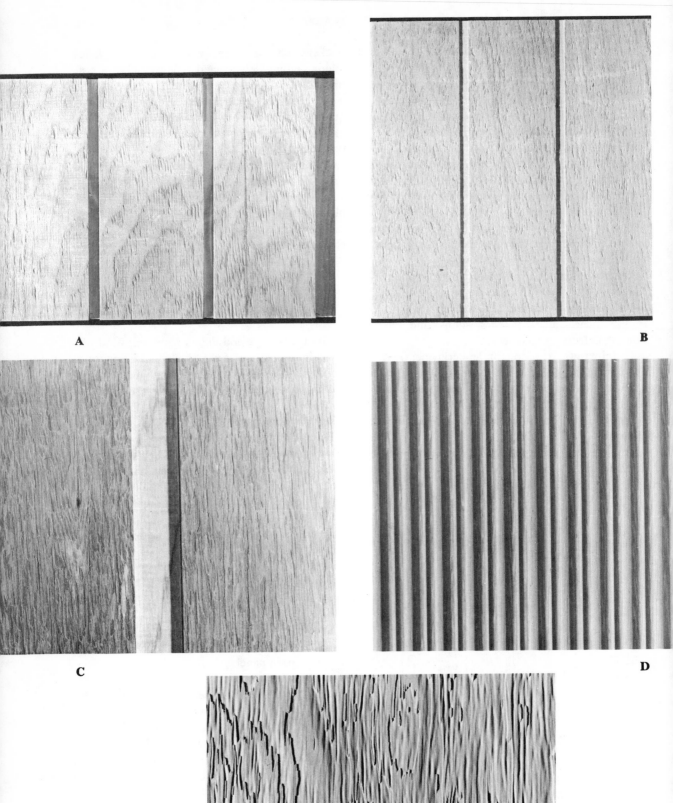

A

B

C

D

E

F

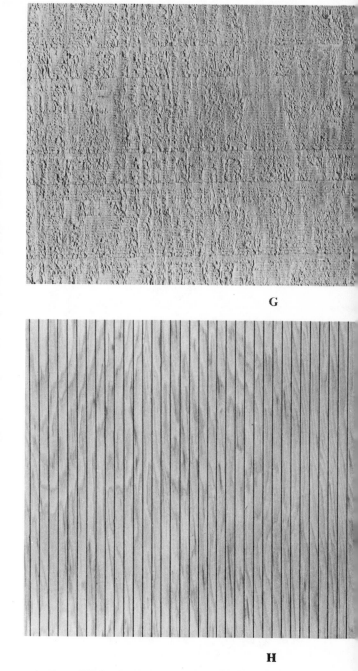

G

H

Fig. 92. A variety of different surfaces have been made available in exterior plywood siding. Photos show 8 of them: (A) Texture 1-11; (B) channel groove; (C) reverse board-and-batten; (D) striated; (E) brushed; (F) kerfed; (G) rough-sawed; and (H) fine line. Photos courtesy American Plywood Assoc.

material, allowing stains, discolorations, light knots, splits not exceeding $\frac{3}{16}$ inch, but no forms of decay.

Reject grade **(Grade 4)** is the same as utility grade, except that the defects are bigger.

Three types of adhesive are used in hardwood ply.

Type I is waterproof and is used for outdoor and marine jobs. **Type II** is highly water-resistant and will retain most of its strength if occasionally wetted. **Type III** is dry bond and cannot be subjected to water, dampness, or high humidity.

Type II is recommended for interior uses, while Type I is used for exterior applications.

Shop for plywood with the following in mind:

For a paint finish, *Sound* or *B*-grade plywood is adequate. However, if you want a natural-wood effect, you must get *good* or *A*-grade face. In both cases, use *utilit*y grade for backs and concealed areas.

SPECIAL NOTE

The amateur carpenter should buy his lumber from a reputable materials dealer. A high percentage of professional builders buy from the same locally established merchants. You can judge a retail yard by the variety of stock it carries, by the knowledge of its attendants, and by their interest in you and your needs. Its personnel should be able to help you pick suitable grades and species, and estimate quantities needed. Of course, you should come prepared with information on the kind of project you plan, its measurements, and how you want it to look.

CHAPTER SIX

Carpentry Materials

The number of building materials available to today's carpenter has increased dramatically in the past several decades. At one time a carpenter dealt exclusively with wood. Now, in addition to ordinary wood strips and framing timber, he can work with plywood, prefinished paneling, gypsum board, hardboard, particle board, and plastic laminate, among other new materials. No longer a simple commodity directly from nature, wood has become a material converted into products designed for specific uses—a versatile material that has been engineered by chemists and scientists into brand-new forms that are better than the old.

The amateur carpenter should be familiar with all of them, because each wood-type material is designed to be used as a complement to or a substitute for wood. Each has particular advantages and limitations; each can be installed with conventional fasteners; each provides a specific type of construction performance.

The secret of wood's new usefulness is the addition of plastic mixes not only in surfacing finishes, but in finishes over outdoor siding, in adhesives used to laminate wood for strength, and in insulation materials in sheet-board form and in batts and blankets like fiber glass. In fact, the use of plastic with wood is so common now that it is apt to go unnoticed because of its ability to blend in with wood and serve successfully in many different guises.

In effect, that is the advantage of plastic. The product of a pretrochemical base, plastic material can be engineered to provide any characteristic desired. For example, it can be structured to be heat-resistant, weatherproof, elastic, and durable, and it can be molded into unlimited shapes.

All carpentry materials that contain plastic have been engineered for easy workability and ease of installation with the do-it-yourselfer in mind. The popularity of these materials with non-professionals is proof of their adaptability. Among the many other advantages of plastic materials are slow aging, easy maintenance, pleasant appearance, controlled texture, fast color, translucency, good acoustical properties, heat insulation, fire control, and dimensional stability.

Prefinished panel surfacing was the first step in the combining of wood and plastic, a step that provided an extremely durable finish to exterior siding guaranteeing a maintenance-free surface for at least fifteen years. Pure plastics are now used in nylon shutters and storm windows, do-it-yourself plastic molding, in downspouts, in siding, and in other materials. Plastic material in building construction alone makes up at least 18 percent of all plastics marketed, according to reports from the Society of the Plastics Industry, an association of firms involved in the manufacture of plastic products.

Whether the material is paint, adhesive, siding, or insulation, it is well within the abilities of the amateur carpenter to use it to great advantage.

Much wood/plastic material can be handled with woodworking tools. Certain types of reinforced plastic, however, can be deformed by an ordinary saw. Saber saws, power saws, and combination handsaws come equipped with extra blades, one of which is designated specifically for

Fig. 93. Storm windows made of heavy-gauge, triple-thick vinyl plus tough extrusion plastic frames can be assembled and installed easily without the use of special tools. Fiber-glass screening is available to replace the vinyl during hot-weather periods. Photo courtesy Tenneco Chemicals.

use with more delicate plastic materials. Other forms of plastic should be cut only with a coping saw, or fine-toothed hacksaw.

Ordinary prefinished interior panel, however, can be handled exactly like a piece of plywood of the same dimension. So can other types of reconstituted wood products like hardboards and particle boards.

The use of plastic substitutes for difficult-to-install materials—like pipes, plumbing fixtures, and electric fixtures—allows the amateur carpenter to accomplish much more successful repair jobs. For example, plastic trim to replace broken molding goes on easier than wood, is main-

tenance-free after application, will not split, and tends to take the proper shape in faultily structured wall joints. That is just one of many such shortcuts the home carpenter can take with the use of plastics in building.

SOLID-WOOD PANELING

Solid-wood paneling is the oldest and most conventional wall cover used in interior decorating. Live wood provides warmth and a cheery feeling that is difficult to approximate with paint or wallpaper. An infinite variety of grain patterns, directly from nature, enhances all types of

Fig. 94. Photograph shows interior of home that has been decorated with strips of redwood paneling in vertical application. Each piece is chamfered, or beveled on both borders, to make a V-groove between pieces to accentuate the joint. Photo courtesy California Redwood Assoc.

wood from plain pine to elegant polished hardwood.

Wood also has a number of natural assets. Age increases its beauty and adds to its character. Besides that, wood is a natural insulator against temperature and noise.

Solid interior paneling comes in thicknesses of ⅜ to ¾ inch, and in widths of 3 to 12 inches. Most of it is sold in 8-foot lengths for quick vertical installation in the average 8-foot-high room. Larger lengths can be ordered from your lumberyard.

Most paneling is jointed for tongue-and-groove application, but some is jointed for shiplap, some is beveled for bungalow siding, and some comes in square-edged board strips.

Exterior types of siding, like beveled siding, can be used inside the home as well, often on a single wall as a contrast to plain paint or wallpaper on the other three. Various styles of application like board-and-batten and channel groove can be used for special effects. The variations are endless. All siding can be applied either vertically or horizontally.

The main disadvantage of solid paneling is its cost. The hardwoods—walnut, oak, cherry, mahogany—are very expensive, especially when purchased prefinished. Even unfinished, these hard-top process hardwoods are costly. Softwood panels of redwood, knotty pine, and western red cedar cost less, when available, but are still not cheap.

In addition to the above, other types of siding that can be used inside the house are eastern white pine, western white pine, sugar pine, and cypress, which are very good quality, with western hemlock, ponderosa pine, spruce, and yellow poplar of fair to good quality, and Douglas fir, western larch, and southern yellow pine of fair quality.

In working with solid paneling, use ordinary carpenter's tools. A power saw will speed up cutting, but is not necessary. Cut solid paneling strips face up with a handsaw or table saw, and face down with a circular or saber saw. Use a fine-tooth blade in any case. You can cut the edges of boards at a slight angle to make them easier to fit into place. The angle should slope downward toward the board's back side.

ATTACHING PANELING IN PLACE

When fastening paneling in place, be sure there is enough studding support at the sides, top, and bottom to keep the boards from bending when heavy pressure is applied to them or when warping tends to twist them out of plumb.

On uneven walls or for situations where a wall is unfinished or only roughed out, furring may be necessary to back up solid panel pieces (see discussion of furring in Chapter Seven).

Apply paneling with regular finish nails, countersink the heads, fill the holes with plastic wood, and sand when dry. If necessary, you can use wood screws for a tighter fit. No predrilling is needed except for certain hardwoods.

Use 6d nails for interior applications. For tongue-and-groove paneling narrower than 6 inches wide, blind-nail through the tongue at every backing support—that is, nail diagonally into the tongue as shown in Fig. 95). For any other type of installation, face-nail twice per stud for boards up to 6 inches, and three times for wider. Countersink finishing nails, fill holes with plastic wood or putty, and sand when dry.

Solid paneling can be applied vertically, horizontally, or even diagonally halfway up the wall to obtain a special wainscoting effect.

For horizontal application, begin at the bottom, and trim boards to fit on the ends. Working from bottom to top, fit each board if the adjoining walls are uneven. If the boards are shorter than the wall's length, join them together only where there is a nailing surface.

Fig. 95. Attach paneling strips to furring by nailing each piece through the tongue diagonally into the backing support. When the next piece is fitted over the tongue, the nail won't be seen.

For vertical application, begin at one corner and fit consecutive boards one after the other. All boards should be 8 feet in length unless floor or ceiling is uneven.

Paneling should be applied directly to wall framing in horizontal applications. If you are adding an interior wall on top of an old one, use furring strips as explained later, particularly if the wall is uneven or if the studs are difficult to locate.

Solid-wood paneling is usually selected for the beauty of its wood grain. However, all wood paneling can take ordinary alkyd, oil-base, or latex paint either clear or pigmented.

PLYWOOD WALL SURFACING

Plywood is essentially a building material that is built up of sheets of wood alternating in grain direction and laminated permanently together with waterproof adhesive. Plywood always has an odd number of sheets since the grain of the face and the back always run the same way. With the grains giving strength to both lateral directions, plywood is a structural material designed and fashioned for maximum strength.

Plywood is used extensively in underlayment in flooring, in sheathing for exterior walls, and in sheathing for roofing. The amateur carpenter will probably not use it for heavy structural work, but he should know how to apply prefinished interior plywood paneling, and how to use plywood for closet interior surfacing and for other projects around the house.

The major event in the do-it-yourself field recently was the development of prefinished plywood paneling. Over 75 percent of all interior paneling is finished at the plant today. This material can be installed over an existing wall and can be used in new construction as well, particularly for add-on rooms. It can be put on directly to an extant wall, to framing studs, or to furring strips over masonry.

Prefinished paneling comes ready for installation, usually in ¼-inch thickness, in sheets 4 feet by 8, 9, 10, or 12 feet. The panel facing is bonded directly to the plywood base as a veneer. The veneer is finished at the factory with permanent seal.

Prefinished surfacing is practically maintenance-free. Once up, the surfacing is there virtually for life. The finishing is usually protected by a film of plastic seal, protecting the clear pigmentation from showing signs of wear and tear for a long time.

Prefinished paneling comes in a variety of face veneers and textures. Its advantages are the advantages of any plywood panel: it is non-splitting, resistant to warping, and very easy to install. Plywood has fair insulating and sound-absorbing qualities as well. And it provides structural rigidity.

Many fine and beautifully grained hardwoods are available as facing veneers. Most of the panels are 4- by 8-feet modules, but prefinished paneling can be obtained in widths from 16 inches to 4 feet. A 16-inch width will just fit across the average studding centers in conventional house framing.

In addition, there are a number of textured softwood surfaces that are not prefinished, but are available with special types of wood surfaces: rough-sawed, brushed, fine-line, striated, texture 1-11, reverse board-and-batten, channel groove, and Medium Density Overlaid for good paint base (Fig. 92). While designed for exterior use, these surfaces can be used inside the house.

INSTALLING PLYWOOD SURFACING

Plywood surfacing that is ¼ inch thick is the most popular size for interior wall work, and is installed by one of four methods:

1 Textured plywood—not prefinished—can be nailed directly to an extant wall through the surfacing into the studs.

2 Prefinished paneling can be installed by nails or by adhesive available in cartridge sizes for use in caulking guns.

3 Paneling can be applied by a clip system that attaches invisibly to walls or studs.

4 Random-width V grooves are arranged in some prefinished paneling so that there is a groove every 16 inches, providing recessed sites for nailheads which won't mar the panel's finish.

Detailed instructions on installing plywood paneling are included in Chapter Seven.

HARDBOARD WALL SURFACING

While plywood panels are made up of sheets of wood fastened together for maximum strength, hardboard panels are made of natural wood fibers that are reconstituted industrially and compressed into smooth grainless sheets designed to maintain rigidity.

They come in conventional modules—4 by 8 feet, and so on—and can be obtained in smaller or larger sizes by special order. They come either in ⅛- or ¼-inch thicknesses.

Several styles are marketed, but the most common is smooth on one side with a rough, wafflelike texture on the other. More expensive types feature smooth surfaces on both sides. Some even have decorative finishes—grooved, channeled, or embossed with designs.

In addition, a type of plastic-coated hardboard panel comes in various colors, which are fixed for permanency. It can be purchased in block colors or in patterns. A simulated wood grain, which looks very much like real hardwood, is the most popular of the specialized surfacings.

The advantage of hardboard is that it has no grain that can become subject to the usual wood defects: rising, warping, checking, knotting, skips, bark pockets, and so on. Since it has equal strength in all directions, it can be used where any hard, smooth surface is wanted. It is excellent for covering tables, counter tops, and desks, and can be used for small shelves in cabinets and for drawer bottoms. It has high strength, durability, and moisture resistance, in addition to its other properties.

Hardboard comes in two grades, standard and tempered. Standard hardboard is used for normal indoor applications. Tempered has a harder surface and is specially treated to withstand moisture. Tempered can be used for all outdoor projects, and for any indoor project that is subject to a great deal of high humidity or moisture.

Since it has no knots or imperfections, hardboard can be used for modernizing almost any wood surface that has gone to pieces. For example, you can attach it to an old door surface to convert it to a new flush door. You can also use hardboard to give a brand-new paint surface to a badly battered or deteriorated gypsum or plaster wall.

A special type of hardboard has perforations at regular intervals throughout the panel. Fixtures in a wide range of different shapes can be used with these holes to create a wall-storage area. Kitchen utensils, tools, small shelf brackets, coathooks, and an endless variety of items can be hung for on-the-wall storage. The perforated panels come in both ⅛- and ¼-inch thicknesses, with the thicker board for heavier objects where additional strength is needed.

Hardboard installation does not require special tools. Use ordinary woodworking tools, and handle them exactly the way you would working any natural wood piece. For example, to fasten sheets on a wall, use nails, screws, or adhesive, or combinations. You can fasten panels to framing members, to furring strips, or to masonry walls in the conventional manner.

In order to prevent nails from popping out later, it is best to use ring-grooved or screw-type nails. Wood screws will give you maximum holding power. Ordinary panel adhesive will fasten hardboard to walls in most situations.

Hardboard is not designed to be butt-ended together like pieces of wood. If you want to join two sheets of hardboard at right angles, attach each to a piece of wood used as a framing member.

The beauty of hardboard surface is that it can be painted with almost any kind of semigloss paint or enamel.

GYPSUM-BOARD WALL SURFACING

By far the most common wall and ceiling material in use in the home today is gypsum wallboard. It can be used both as a finish material or as an underlay for another surface—both inside and out. Its widespread use is due to its relatively inexpensive price, compared to other types of material that afford a similar paint-finish surface.

Gypsum wallboard is called "dry wall" to distinguish it from "wet wall," a term once used to describe a plastered wall. Gypsum duplicates plaster and is molded into a rigid sheet by cardboard surfaces that hold the inert gypsum rigidly in a firm structural sandwich. It has the same

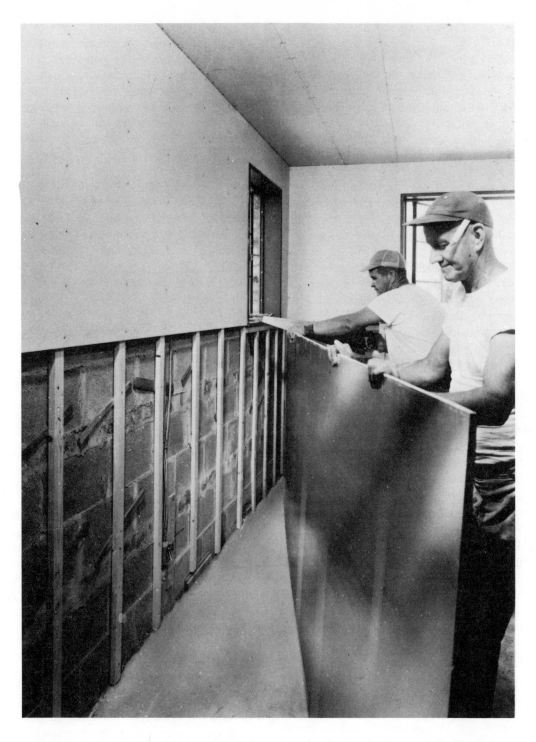

Fig. 96. Typical application of gypsum board to the bottom half of a wall. Vertical strips fastened to concrete-block wall are furring strips, used in place of regular studs for attachment. Gypsum board is fastened directly to these strips. Photo courtesy U. S. Gypsum.

properties as a plaster wall and can be installed more easily.

Two main types are available. One is the standard plain wall panel that can serve as a complete, finished wall or as an underlay. There is also a predecorated variety that can be installed as an attractive finished wall. With this type, a prefinished plastic surfacing is bonded to the gypsum board in a variety of designs that are both colored and decorated.

Widths are 2 or 4 feet, with lengths running from 8 to 16 feet. There are several thicknesses: ⅜ inch as regular underlay; ½ inch for final wall cover; and ⅝ inch for walls between garage and living structure.

Gypsum-board panels are designed with different types of joining seams. With decorative and finished wall types, the edge seams are made to be shown on the wall. They may also be taped and cemented to conceal joints if desired. Ordinary underlay panels are made with joints designed to be taped and spackled for concealment.

The amateur carpenter will find that taping and spackling gypsum joints can be quite a demanding job. It is, however, described in detail in Chapter Seven.

Most gypsum board is applied to the wall with special dry-wall nails.

In cutting gypsum board, use a utility knife and straight edge, scoring along the cutting line with the knife until the paper surface is cut through to the gypsum. Place the board over an edging of some kind—a 2 by 4, or the edge of a large worktable—right along the cutting line, and snap the excess piece off. It will break along the scored cut. Then slice the paper along the backside of the board.

To cut openings for wall fixtures, use a saber saw or a compass saw after drilling a starting hole, or score the cutout pattern on both sides of the board with a sharp utility knife and punch it out with a hammer.

To fasten gypsum board, all four edges must be supported on studs horizontally and vertically. Studs can be set 16 inches on center or 24 inches on center. For horizontal installation on an 8-foot stud wall, fasten one row of fire-blocking—horizontal stops halfway up the studding—to serve as a nailing base.

With predecorated gypsum board, use wall-board adhesive to fasten the sheets to studs or wall, and then nail around the perimeters with color-matched nails. Do not break the surface where you can't cover it with molding.

Apply regular gypsum board horizontally, working from a top corner down and across. Apply corner moldings, and conceal the joints and nail dents after installation.

Apply predecorated gypsum board vertically, beginning at one corner and working across the wall. Most decorated gypsum board requires the application of underlay to support it. When the decorated board is installed, apply corner moldings if needed.

Detailed instructions and illustrations are included in Chapter Seven. For the type of trim to use, see the section on millwork and moldings in the same chapter.

PARTICLE-BOARD WALL SURFACING

Particle board is an engineered, grainless wood panel product, usually intended for applications where smooth surfaces and uniform stability are important. The most common application around the house is underlayment for flooring.

Although the amateur carpenter is not likely to find himself using particle board, since it is the province of the professional, he should recognize it and understand its advantages.

Several new types of particle board now include a style "filled" for easy painting, a hardwood-veneered particle board for paneling, one toxic-treated to resist fungi and insects, a moisture-resistant type treated with phenolic, a phenolic-overlaid exterior type, an acrylic-overlaid kind for weather- and abrasion-resistance, and a fire-retardant type.

OTHER WALL-SURFACING MATERIALS

In addition to the main types of building material already discussed, there are a number of other types with which the amateur carpenter should be conversant. These include insulating wallboard, fiber-glass paneling, cork tile, real-brick veneer, real-stone veneer, plastic laminate, compressed wood paneling, and wood shingles and shakes.

Insulating wallboard resembles hardboard cov-

A

B

C

Fig. 97. (A) Panel of brick veneer being fitted in pattern. (B) Each panel is nailed into place or fastened with adhesive. Mortar is applied (C) with caulking gun for finishing touch, covering nailheads and cut edges and corners. Photos courtesy Masonite Corp.

ering, although its properties are slightly different. The material is lightweight and easy to handle, and comes with textured surfaces in many different prefinished colors and designs.

Because of its insulating properties, it can be used as a substitute for other types of insulation. The material that makes this board favorable for insulation also makes it easy prey to marring and scratching.

Insulating wallboard is used most successfully in areas where walls are rarely subject to puncturing shocks. You should plan to apply insulation board to a hard surface that will hold it up. The board can be applied over furring strips, or directly onto rough framing.

Use nails or staples to secure the panels to the wall.

Fiber-glass paneling is also available, usually for outdoor installation where its translucent properties make it particularly effective. The paneling comes in thin sheets that are either flat or are corrugated like sheet metal.

Paneling can be used as wall covering, as an actual wall, or as a divider or partition for larger areas. Fiber-glass paneling is rigid and needs support only at top and bottom.

Polystyrene paneling is similar to fiber-glass paneling, but is designed only for interior installation. A type of polystyrene paneling creates a stained-glass effect that is very provocative.

A most unusual wall surfacing can be obtained by using **cork tile.** These tiles come in 6- by 12-, 12- by 24-, and 24- by 48-inch rectangles, with three thicknesses available from $\frac{1}{8}$ inch to $\frac{5}{16}$ inch. A wide range of tones from light to dark brown is obtainable. Cork tile must be installed by adhesive, and it can be applied only to a solid wall.

Real-brick veneer is available for use in the interior of the home, particularly for fireplace installation or single walls, to add special accents to finished rooms. The brick is actually the real thing, laid in the manner of an ordinary brick wall. However, the front inch or so of the brick facing is all that is used to provide the façade. The facing can be attached to gypsum board, plaster wall, wood paneling, or plywood with adhesive or special fasteners. Many varieties of real brick, with different styles and designs, are on the market.

Real-stone veneer is also available for interior use as well as exterior use. These are real stones which, like real brick, are cut off to provide a façade around fireplaces or on regular walls as interesting visual accents. The stone veneer is attached either with special adhesive or fasteners.

Plastic sheets called **laminate** can be attached to walls for interesting textural surfacings, or just to provide a perfect surface for paint. Laminate is applied with adhesive made specially for it. It is about $\frac{1}{8}$ inch thick, and is a plastic formulation with a rugged, durable, and unchanging surface. It is virtually indestructible under ordinary circumstances.

Compressed paneling is a type of reconstituted wood-flake-and-chip combination bonded together with special synthetic resin. This compressed plastic paneling comes in 4- by 8-foot sheets and thicknesses of $\frac{3}{8}$ to $\frac{3}{4}$ of an inch. This material takes interior paint of all kinds as well as all kinds of wood finishes. Install it just like rigid plywood paneling.

Wood shingles and shakes can also be used as surfacing for an interior wall, although they are usually used for exterior walls or roofing surfaces. For installation, see Chapter Eleven.

Section III

INSIDE WORK

The home carpenter's work is distributed about evenly between inside the house and outside the house. Inside work generally involves a number of categories, arbitrarily divided in this book into four chapters. They include Chapter Seven on walls and wall surfacings; Chapter Eight on floors; Chapter Nine on doors and windows; and Chapter Ten on built-ins and added rooms.

The discussion of wall work in Chapter Seven describes the general structure of the interior wall of a wood-frame house, and tells how to attach the various types of surfaces. It also includes details on wall-surface repair.

Chapter Eight on floors concentrates on wooden floors, although it includes a section on laying and repairing resilient tile. Ceramic tile is not included, nor is concrete work.

Chapter Nine on doors and windows includes the anatomy of both, and the step-by-step procedure in hanging a door and in readjusting one that sags or binds. Procedures for repairing glass windows are also included.

Because shelving and closet space are so important to any homeowner, a working knowledge of building drawers and installing shelves is a must for any amateur carpenter. Chapter Ten has a special section on finishing off a cellar or attic space.

CHAPTER SEVEN

Interior Walls

BEARING AND NON-BEARING WALLS

The ability to differentiate between a bearing and a non-bearing wall is essential before beginning any carpentry project on an interior house wall.

Exterior walls that run at right angles to ceiling rafters and floor joists and the interior walls running parallel to these exterior walls are called bearing walls because they help bear the weight of the roof and/or second story.

Cutting into, altering, or removing any part of a bearing wall may seriously endanger the entire structure of the house. Be sure not to plan major changes in bearing walls without consulting an architect, engineer, or contractor. It's best to hire someone to do any job planned on a bearing wall.

The easiest way to determine whether or not an interior wall is bearing or non-bearing is to climb into the attic or crawl space above the ceiling joists and see if the joists are joined over any particular crossbeam or wall. The wall supporting such a beam or supporting the joists *is* a bearing wall. Although the joists may be long enough to span a house's width, their midsections may be resting on an interior wall for added support. That wall must be considered a bearing wall as well.

All other walls in the house—interior and exterior—are non-bearing walls and are used to divide rooms or close in the ends of the structure. Interior non-bearing walls are made of light materials and can be altered and reworked without causing serious structural problems. Jobs on non-bearing walls are the only ones that should be tackled by the amateur carpenter.

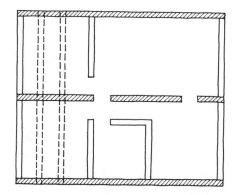

Fig. 98. Shaded walls of house in plan are load-bearing. Note that ceiling joists run at right angles to load-bearing walls. Joists usually break at center wall, spanning only half of house's width.

STRUCTURE OF AN INSIDE WALL

Even though you may never be called upon to build an interior wall, you should know how it is put together in order to do ordinary carpentry repair work on it. The same type of construction, with important modifications, is used not only on inside walls, but on outside walls, floors, ceilings, and roofs as well.

Basically, a wall is a skeletal framework of lumber with covering applied to it. The covering can be wood, masonry, plaster, or other materials such as aluminum or plastic.

The framework is composed of timber called studs. A stud is usually a 2 by 4, although in certain construction situations a 2 by 3 can be used if the wall is non-bearing. A type of framing member called a rafter, usually a 2 by 6 or 2 by

8, is used to form the framework for a ceiling and floor.

In most frame-house construction, studs are placed 16 inches apart—"16 inches on center," meaning that the *centers* of the studs are 16 inches apart. The 16-inch distance accommodates construction materials usually prepared in modules of 4- and 8-foot widths and lengths, with the ends meeting at a fourth stud.

Most walls in homes are built in heights of 8 feet between floor and ceiling, with materials prepared in modules to fit this height. Solid wall-paneling strips can be purchased in 8-foot lengths, for example. Plywood and gypsum board, materials most commonly used as covering for interior walls, come in 4- by 8-foot modules.

The framing studs do not rest on the floor itself, but are nailed to a 2 by 4 strip called a sill or floor plate running across the floor. Its opposite number nailed to the ceiling directly above it is called a plate or top plate. Once these two members are attached to floor and ceiling, two end studs are nailed to opposite walls, either by toenailing or by nailing directly into a stud or underlay timber. With both these end studs in place, the rest of the wall studs are placed at 16-inch intervals along the line of sill and plate, checked for distance with a rule and perpendicularity with a level and toenailed in place at sill and plate with 10d common nails.

An interior wall usually has a door, which can be placed at either end or somewhere in the middle. Most doors are 6 feet 8 inches in height, and of a width from 24 to 36 inches. Upright studs and sill plate are omitted in the portion of the wall where the door will be hung.

To leave room for a door in a wall, mark the width of the door on the sill plate and saw the sill at the marks and remove the excess. Nail double upright studs on each side of the door space, leaving enough room for the casing, or finish pieces that make up the visible part of the door frame. At the top of the door, place two horizontal studs parallel and upright, and nail to the double uprights. These are what are known as double headers. Fasten a cripple stud, a short 2 by 4, between the top plate and the double header to keep the two studs in position.

Horizontal fire stops are attached 48 inches on

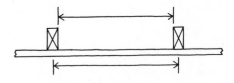

Fig. 99. Diagrammatic sketch shows the placements of studs within a wall, 16 inches on center. Note that the clear span between the two studs is 14½ inches.

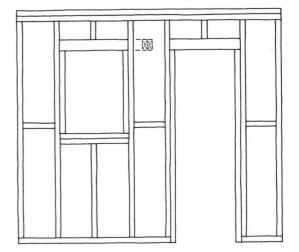

Fig. 100. Sketch of wall framing shows all the principal parts of the wall of a wood-frame house. When double headers, parallel 2 by 4's, are turned to the vertical, a plywood strip is inserted between them to bring their combined thicknesses to the width of a 2 by 4. See inset.

center from the floor between each stud to keep the members as they season from twisting and throwing the finished wall out of line. If the wall surface is to be attached in horizontal modular units, the fire stops must be exactly 48 inches from the floor to provide a nailing base; however, if solid-board paneling is to be used, the fire stop height can be alternated a few inches for ease in nailing.

For a wall surface of gypsum board or plywood paneling—either prefinished paneling or underlayment for a paint or wallpaper surface—nail the modules to the studs as described in Chapter Six. Be sure to apply prefinished paneling vertically and gypsum board and plywood horizontally.

Since most walls, floors, and ceilings are not quite plumb where they join, you may find that the first module does not fit tightly to floor and wall. It is advisable to cut the piece until the module is erected squarely so that adjoining modules will fit cleanly, otherwise the first error will be magnified module by module. Check the horizontal line and the vertical line with a level, making sure the first rectangle is right, and then make whatever alteration cuts are necessary in the piece to make the module fit.

If the module does not quite meet ceiling, wall, or floor, don't worry. Molding will cover the gaps. Make sure the joints *between* the modules on the wall surface are tight, however; they must be perfect, particularly if the module is gypsum board which demands taping and spackling at the seams to receive a finish.

HOW TO USE FURRING

In working with certain types of wall material it is sometimes necessary to provide a solid nailing base in addition to 16-inch on-center studding. Extant walls that are broken, warped, or twisted out of line may also require such a solid nailing base. So does a wall in which the studs are difficult if not impossible to locate because they have been placed at random intervals upon installation. It is also true of the problem arising in adding wood paneling or even plywood to a masonry wall.

To provide such a solid nailing base, 1-inch members of material called *furring strips* are used. A furring strip is simply a foundation member attached to studs, wall, or masonry surface, to which a wood, plywood, or gypsum-board surface can be nailed (Fig. 101).

Usually a 1-inch-thick strip from 2 to 4 inches wide, the furring member is attached by nails to studs and to studding through dry walls. When applying furring to masonry walls, use masonry nails, or expansion or toggle bolts as explained in Chapter Three.

Furring generally runs at right angles to the material installed over it. In the case of plywood and gypsum board, it runs at right angles to the module length. However, in all cases, furring should be provided all around the perimeter of the module.

To provide furring for vertical paneling strips, fasten one strip along the top plate and one along the sill plate. Then run a third furring strip 32 inches from the ceiling, and a fourth 32 inches from the floor. This provides four horizontal running strips equidistant from one another on which to attach paneling members. A vertical furring piece should also be nailed to each end of the wall, and along the door frame all the way to the ceiling to give the paneling a solid border.

Obviously, it is not necessary to supply furring for horizontal paneling strips applied directly to wall studs, since the boards can be attached directly.

In the case of furring strips for a wainscot effect—paneling only about halfway up the wall—use horizontal furring strips as described, using one to anchor the top of the wainscot. Add another next to it to provide a nailing base for the bottom border of the dry wall where it meets the wainscot.

If furring is unnecessary, don't bother with it. The 1-inch strips simply take up space. If for any reason whatsoever you can't get a good purchase on any wall surface or on studs through a wall, by all means use furring to provide a proper nailing base.

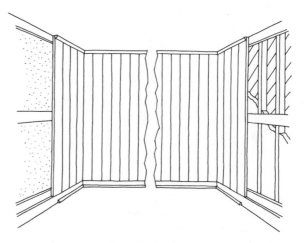

Fig. 101. Illustration shows how furring strips are placed to receive paneling. Note that the center strip is midway between top and bottom. For a room with a 4-inch baseboard, use 1 by 8 furring strips as indicated. Picture shows furring on concrete wall and on ordinary wood-frame studding. Furring is attached to concrete by anchor screws.

ATTACHING SOLID-BOARD PANELING

For solid-board paneling applied horizontally, attach panels to the studding directly. Begin at the bottom, fitting the bottom piece and making sure it is exactly level before fastening. If the piece is not level, cut it to fit the floor until the top side is level. Make sure the ends that abut the adjoining wall or walls fit correctly. It is frequently necessary to jockey the first piece of paneling in place so that the rest of the pieces will align for perfect fits.

You can usually apply horizontal panel strips so that each run is one piece of wood. However, if the wall is too long for that, cut the pieces so that they meet exactly in the center of a stud. Make sure that adjoining runs do not meet on the same stud.

In the case of vertical-board paneling—and diagonal paneling as well—you will need to use furring pieces as described above.

With the furring in place, attach the first solid board at either the left or right wall, making sure it fits exactly vertical to the floor. Check its position with a level. Now fit the second piece to the first, check it with a level, and attach it.

Moving from one piece to another, complete the wall. Ripsaw the last panel that goes against

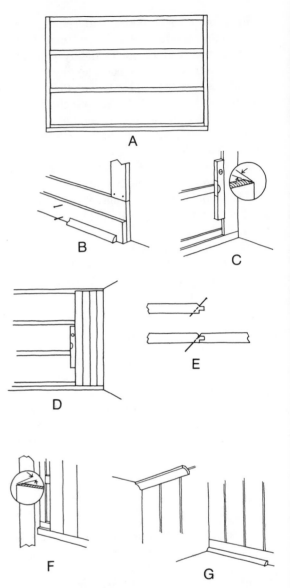

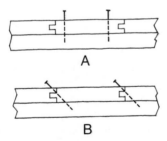

Fig. 102. Tongue-and-groove paneling can be applied either through face-nailing or blind-nailing. (A) Shows how a strip of tongue-and-groove is face-nailed to sheathing or underlayment. Two nails are needed for each joist. (B) Shows how to blind-nail a strip of tongue-and-groove. Only board ⅟1 is in place when nail is driven in at a 45-degree angle through the top of the tongue. Later board ⅟2 is pushed up against tongue, and it, in turn, is blind-nailed in similar fashion.

Fig. 103. Drawings show how to install vertical paneling on an interior wall. Install furring strips as shown in (A). Note that the furring strip along the floor is a 1 by 8, and the rest are 1 by 4s (B). Trim all paneling, baseboards, and molding to fit. Fasten baseboards to the 1 by 8, face-nail, countersink, and fill with plastic wood. Trim first panel (C) and install it tightly in the right-hand corner, butting down on the baseboard. Check first several panels for vertical plumb, adjusting slightly as each panel is added (D). Blind-nail all paneling to furring strips (E). Trim the last panel (F) and slip into place after ripping it if needed. Nail. Attach cove or crown moldings, base shoe at floor contact, and, if needed, a quarter-round trim in the corners (G).

Fig. 104. Schematic drawing shows how paneling is applied in horizontal strips. Each board should end at a stud to establish firm backing. Note that no two adjacent horizontal runs break at the same stud.

Fig. 105. Photograph shows method of checking piece of paneling for true vertical. Scrap piece of tongue-and-groove protects tongue as handyman taps it into place. Note chisel used as shim to keep board tight to top border. Floor molding will cover discrepancies at bottom while top joints are in plain view. Photo courtesy Western Wood Products Assoc.

the door or wall so that it fits exactly, or near enough to be covered by the door casing or corner molding.

APPLYING A NEW WALL TO AN OLD WALL

If you are applying a new covering to an extant wall, nail gypsum board, paneling, or prefinished panel modules directly through the old wall into the studs, or furring, making sure the nails penetrate the studs at least 1 inch.

It is sometimes quite difficult to locate studs in a wall you have not built yourself. You have to assume that the studs are set 16 inches on center —16 inches apart. There will always be one stud at the beginning of a wall at the corner to support the wall surface. Measuring 16 inches out from that, you should find a second stud. However, this is only true if you start at the same side of the room as the original carpenter. Locating the second stud is the important part of the job.

If the second stud is not 16 inches from the

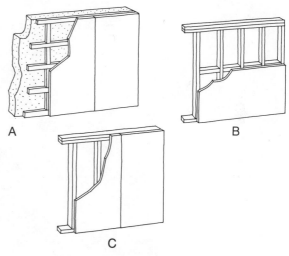

Fig. 106. (A) How to install sheets of plywood paneling in vertical position over an extant masonry wall. First install 1 by 2 furring strips with expansion bolts, concrete nails, or explosive fasteners. Block all unsupported edges, as indicated in the drawing, with short vertical furring strips. (B) How to install plywood module horizontally to studding. A fire stop should be built into the middle of the wall horizontally to support the edges of the paneling. (C) How to install plywood vertically directly to studding.

A

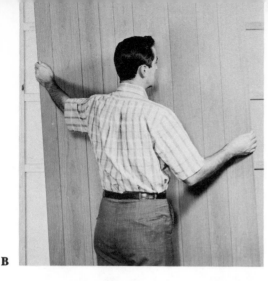

B

C

D

E

F

Fig. 107. Hardboard paneling is attached like regular plywood paneling. (A) Nail 1 by 2 furring strips to the existing wall at stud locations, spaced 16 inches apart. Apply vertical furring where panel edges are to be bonded. Furring can be attached vertically with adhesive if studs are hard to reach. When furring strips are in place, apply adhesive to all of them in long beads in the center of each member. Then move panel into position (B) over the furring strips and press into place. With uniform hand pressure (C) press panels firmly into contact with the adhesive bead. Then install two nails at the top of the panel to maintain its position (D), leaving heads exposed for easy removal later. Continue bonding additional panels as outlined in the first steps. After 15 or 20 minutes, reapply pressure, using a padded block of wood and a hammer or mallet to all areas to be bonded to provided a final set (E). Then carefully remove the nails by protecting panel surface with a scrap of carpeting (F). Photos courtesy Masonite Corp.

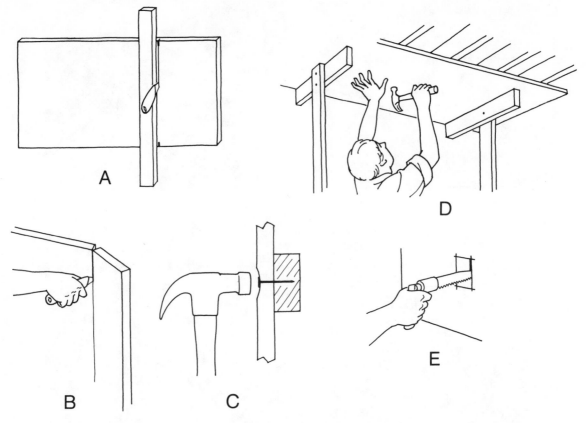

Fig. 108. To install gypsum board, first place the panel with the good side up, and measure and mark panel at opposite edges. Line up the points with a straightedge, and hold straightedge firmly against the board while scoring the paper through to the gypsum with a utility knife. Hold the knife at a slight angle away from the straightedge to prevent cutting into the board edge (A). Now break the core of the gypsum board (B), snapping away from the scored face paper. Run the knife through the back paper to separate it. Use annular-ring nails (C) for attaching gypsum board to studs or furring pieces. Space nails 7 inches apart on ceilings, 8 inches apart on walls, and at least ⅜ inch from ends and edges of panels. Use 1¼-inch nails for ¼-, ⅜-, and ½-inch-thick panels. Use 1⅜-inch nails for ⅝-inch panels. Hold the panel tight against the framing and nail the center of the panel first, outside last. Leave a small dimple at the nailhead for filling later with spackling. Do not overdrive or countersink nails. Ceilings should be applied first (D). Two people should handle the job, but if you're alone, make a T-brace as shown, a half inch longer than floor-to-ceiling height. Wedge the brace between floor and panel to support it while you nail. Measure the locations and sizes of all openings in the panels and make cuts with keyhole saw (E). Fixture plate must cover cutout completely.

corner, test the wall with your fists by knocking on it. Gypsum board will usually give a hollow sound if you pound on a portion between studs. When you pound directly over a stud, you will hear a hard solid sound. Drive a test nail in at the supposed location of the stud. It will grab wood immediately after it has passed through the gypsum board. If it does not, try again somewhere else.

Once the second stud is located, you can usually measure off 16-inch sections to catch all the studs with fair accuracy—unless one or two have warped. Using small nails to test and locate, mark the studs on the wall.

Set the first 4- by 8-foot module so that the end hits the stud directly in the middle. If necessary, cut a bit off the end if it tends to overlap the stud. You'll need the stud to attach the end of the next module.

If the module is not exactly at right angles to the floor, cut it to fit and check both side and top with a level before attaching. Then drive nails in through the studs at the edges and intermediately. Attach top and bottom of the module to the sill and top plate.

At the door, cut the module to fit at the double stud from floor to ceiling. Begin the next module on the other side of the door at the double stud. Later, cut a piece out to fit above the door and attach it after both side modules are nailed in place.

HOW TO USE TRIM

The finish wood around doors, windows, and at ceiling and wall joints is called millwork, because it is extensively "milled" to its form. Millwork is designed specifically to "trim" up the appearance of a wall joint that does not fit quite properly, and is thus called trim. In all walls of a wood-frame house there is trim called baseboard molding along the joint between the floor and the wall, cornice molding along the joint between the ceiling and the wall, and corner molding on certain wall corners that present structural problems.

There are five basic kinds of molding:

Casing is the most general type of trim and is used in every room of the house. It completes the trimming of doors, windows, and other openings. Casing may also be used to trim cabinets and provide certain decorative effects. The groove, shown in the drawings, enables the molding to fit snugly on any wall surface.

Baseboard is also used in every room of the house. It runs along the bottom of the wall where it joins the floor surface. It protects the wall bottom from everyday wear and tear, including shoe marks and vacuum cleaner blows.

Cornice is also used in every room of the house. It softens the harsh lines where two planes meet, usually at the meeting of wall and ceiling. The cornice stretches across two right-angle surfaces, usually in a diagonal line. It is also used to

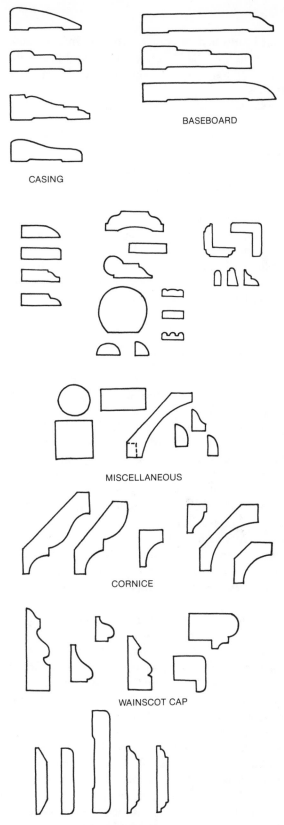

BASEBOARD

CASING

MISCELLANEOUS

CORNICE

WAINSCOT CAP

CHAIRRAIL

Fig. 109. Types of moldings, all shown in cross-section.

trim beams and exterior overhangs. In combination with other moldings, it can be used to decorate mantels, picture windows, and painting frames.

Wainscot cap is used in any room that has wainscot paneling; that is, paneling reaching up to about waist height. Some wainscot-cap patterns have a wraparound tip that is used to conceal defects in craftsmanship. The same molding can be used as a fancy cap for ordinary baseboard.

Chair rail is used for dining rooms, playrooms, and other areas where chair backs are moved constantly and may scratch wall surfaces. The molding should be placed at the proper height for the actual chairs in use. Note that modern chairs tend to be several inches lower down than old-fashioned chair backs.

Miscellaneous trim includes *quarter rounds,* also called *base-shoe molding* when used in conjunction with other molding; *half rounds,* used for door and window screens; *door and window stops,* used in door and window casing; *shelf cleats,* used for wooden-shelf supports; *battens,* used for board-and-batten combinations; *lattice,* used for outdoor garden installations; *picture molding,* used for framing paintings and photographs; *hand rails,* used for stairs; *screen molds,* used for window and door screening; *glass beads,* used in window installations; *closet rods,* used in clothes closets; *screen stock,* used to make window and door screen frames; *workshop stock,* used for various types of supports and blocking needs; *picture framing,* used for paintings; and *baseboard trim,* used in a variety of baseboard situations.

Casing will be explained further on in Chapter Ten on doors and windows.

Baseboard is the largest type of molding, sometimes coming in 1- by 4-inch strips with plain or fancy tops. To install, nail floor molding directly to the wall studs through the surface with casing nails, countersink, cover with putty or spackling, sand, and paint.

To attach the *cornice,* fasten it at the joint of wall and ceiling with a casing nail driven directly in at a diagonal. Countersink the nail head, cover with putty or spackle, sand, and paint.

Most interior corners can be taped and spackled if gypsum board is used. However, for panel-ing—particularly prefinished paneling—*quarter rounds* can be used to hide any mismatches or gaps in the edges of the paneling. Apply with finish nails.

For inside corners, where baseboard and cornice meet, miter the ends of the molding pieces. Mitering is explained in Chapter Two.

In dealing with cornice, it is a tricky business to make a 45-degree angle cut across a piece of wood that must be applied at a diagonal. Hold the cornice piece up to the ceiling corner exactly as it is to be nailed in place. Mark the miter cut as you hold it there. Then place the cornice strip in the miter box so that the ceiling edge is on the flat bottom of the box and the wall edge is against the proper side. Then make the 45-degree cut.

SPACKLING AND TAPING GYPSUM BOARD

If you are dealing with board strips, plywood, or prefinished paneling, your job is all done once you have the material in place, except for attaching the various molding members. However, if you have applied gypsum board, it is necessary to spackle and tape the seams to prepare the surface for painting.

Materials needed are tape and spackling compound, plus two or three sizes of putty knives. The materials can be bought at any building-supply house or paint store.

The first step in spackling is to make sure all gypsum-board nails are firmly pounded into the studs, the gypsum surface dimpled, with the mark of the hammer head clearly imprinted in the surface. No part of the nail should protrude above the gypsum-board surface.

With a small putty knife, apply spackling compound to the joint until there is no gap between the gypsum-board modules and no sign of any nailhead or dimple.

Allow the spackling compound to settle a moment or two, and then unroll the spackling tape from top to bottom along the line of the spackling compound, holding the top with one hand, and pressing the tape on by running a wide putty knife down the tape as it unrolls.

Make sure the tape adheres to the spackle. If there are gaps, or if the tape bulges out any-

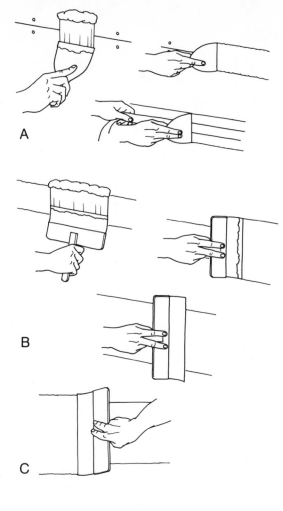

A

B

C

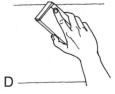

D

Fig. 110. Taping and spackling of joints in gypsum board begins with the application of a large daub of spackle across the joint (A). Draw the spackling knife along the joint, leaving no bare spots. Immediately apply reinforcing tape, running the knife along to press it into the compound. Remove excess compound, but leave enough under the tape for strong bond. Let dry. After at least 24 hours, sand the tape and spackling to a level surface. Then apply spackle as indicated in (B) over the joint and the tape with a larger knife. Extend the spackle 2 inches beyond the taping coat. Feather both edges. Let dry. Apply third coat (C), 2 inches wider than the second coat. Let dry. Using a fine-grade sandpaper wrapped around a sanding block, sand the entire area (D), lightly enough not to rough up face paper. Don't use a power sander.

where, apply spackle until the bulges disappear and the tape sticks to the compound at all points.

Allow the tape to dry in place.

Apply a second coat of spackling compound over the tape, using a wider putty knife and spreading spackle over a wider area so that there is no sign of the point where the first layer ended. Smoothing out the borders is called "feathering."

Let the second coat dry.

Apply a third coat of spackle with as wide an applicator as you have, say, 12 inches or more, letting the compound spread out over 6 to 8 inches from the center of the joint. At this point there should be no indication whatsoever of nail holes, joint, or tape. If there are holes, fill them up and feather the edges.

Let the third coat dry.

Sand. The surface is now ready to accept primer.

REPAIRING PLASTER-WALL SURFACE

One of the most common problems you may have with a plaster wall is filling a settling crack caused by the splitting of the plaster surface as the floor shifts beneath it or as the wall twists slightly with age.

The easiest way to smooth over a small crack is to use spackling compound. Apply spackle with a wide putty knife, smoothing over the edges. Let the first coat dry. Then apply a second coat, feathering the edges farther out. The surface should now be ready for painting.

If the crack is a wide one, you may have to use quick-dry plaster compound, which comes in powder form to be mixed with water. To prepare the crack for the plaster compound, cut in at an angle on each side of the crack, making the crack wider deeper down. The canted angles will then be prepared to hold the compound in place as it dries.

To apply the compound, follow the directions on the container. Mix the dry compound with water to the desired consistency, and apply with a trowel or putty knife, pressing it into the crack. The compound will expand as it dries, filling the crack, and forming a tight fit that will not drop out.

If the compound dries below the surface of the

wall causing an indentation, apply more compound and smooth it over until it dries flush. Let it dry. Sand. The surface is now ready for paint.

REPAIRING GYPSUM-BOARD DENTS

Gypsum board frequently develops holes and dents in its surface. The easiest way to repair such damage is to apply spackling compound to it. Use a putty knife and spread the compound on carefully, allowing it to dry. If the surface is not smooth, apply another coat, smoothing it out and feathering it as described above in the section on spackling and taping.

If the hole is too big, or if the gypsum board is actually broken clean through, cut out the affected area, making a square of easily manageable shape. Cut out a matching piece of gypsum board to fit into the hole.

Apply spackling compound to the inner edge of the broken gypsum-board wall. Cover the edges of the repair piece with compound. With wood screws, attach a strip of wood to the gypsum-board repair piece. Using the strip of wood as a handle, hold the repair piece in the hole as the spackling compound dries.

With the spackle dried enough to hold the piece in place, apply more compound along the line of the break, feathering the edges outward. Let the new layer dry. Apply a third layer of spackling compound, feathering out to 6 or 8 inches. When the final layer dries, the surface is ready for paint.

Remove the screws in the piece of wood. Apply spackling compound to the screw holes in the gypsum-board patch and let it dry. Sand.

REPAIRING BURST SPACKLING TAPE

In some instances, especially in the case of a ceiling in which the spackling tape pops because of flooding upstairs, you will have to remove the tape and apply new.

To remove old spackling tape that has sprung out, use a utility knife and cut along the joint. Peel back the tape, letting paint and old spackle come with it. When the tape is removed, clean out the joint area with a putty knife or paint scraper, removing all old spackle. If the joint is damp, let it dry out completely before going to work on it.

When the joint is dry and ready for taping, apply spackle as you would in starting a new joint. Tape as described. Respackle over the tape when it has dried on. Respackle and feather out to the edges. Let it dry. Sand.

REPAIRING POPPED CEILING NAILS

Settling, excessive vibration, or a variety of types of blows can loosen ceiling nails, making them pop out under the paint surface to form ugly blobs. If the nail is far enough out, remove it with a claw hammer. Drive in a new gypsum-board nail in a position near the old nail hole. Spackle the hole where the old nail was, dimple the new nail, and spackle. Then, after these applications have dried, apply a layer over both and feather out.

FILLING GYPSUM BOARD, DRY WALL, AND WALLBOARD HOLES

When you remove shelves or other pieces that have been screwed into gypsum-board wall surfaces and into studs, ugly holes appear where the screws went through. Cut out excess cardboard and gypsum, and push spackling compound into the holes with a putty knife. Smooth over the holes by feathering out from the center. Allow to dry. Sand. The new surface is ready for painting.

REPAIRING HOLES IN PREFINISHED PANELING

Dents and holes appear in prefinished paneling, even though the surface is designed to take most kinds of blows without evidence. If the holes are big enough, push plastic wood into them with a putty knife, smoothing over as closely as you can. Then, when the plastic wood has dried, sand it carefully, making sure you don't remove any of the prefinished surfacing. Once the plastic wood has been sanded, touch it up with a wax pencil tinted as near as possible to the color of the paneling.

Almost any kind of hole in prefinished paneling can be covered up with plastic wood. If nails pop out, either remove them and fill the hole with plastic wood, or pound them in and cover the hole in the same manner. Sand and touch up with a tinted wax pencil.

REPAIRING A SOLID-BOARD PANEL

Sometimes solid-board paneling splits as it seasons in place. Fill the split with plastic wood, and touch it up when it dries with a properly tinted wax pencil. If the split is ugly and esthetically unpleasing, remove the piece of paneling and replace as follows:

Using a power saw with the blade adjusted for about ⅞-inch cutting depth, run the saw up the middle of the damaged piece of wood, severing it in the middle. At the ends where you can't reach with the power saw, use a keyhole saw. Remove all nails from the panel, top and bottom.

Insert a pry bar into the sawed gap and work the damaged halves out of position, taking care not to split or injure the adjoining panels' tongue-and-grooves. You should be able to pry out both halves without breaking either adjoining panels.

Cut a new piece of paneling to the same size as the damaged one. Readjust the power saw to a little over ¼-inch cutting depth. Carefully remove the *under* or unseen groove of the new piece of paneling, leaving the outer or visible groove untouched.

Insert the new panel tongue first into the grooves of the panel in the wall. Then push the rest in place. The removed groove will allow the new piece to fit right over the tongue of the remaining panel.

Face-nail in place, countersink, apply plastic wood, let dry, sand, and touch up with a wax pencil.

REPLACING MOLDING

Since cornice and baseboard molding is usually always mitered at the inside corners, it sometimes presents difficulty in removal. However, with the proper patience and skill, the job can be done without too much aggravation.

Molding applied by professional carpenters is put up with casing nails, which are long nails like finish nails with slightly different heads. The nails are countersunk and the heads covered with putty before paint is applied.

In removing a cornice molding, insert the point of a small pry bar in between the ceiling and the top of the molding. If you can see where the nails are, use the bar near each nail, jockeying it loose from the wall.

Start in the center of the cornice, away from the corners. Since the corners are mitered and are held in by the natural pressure of the cut, the center of the strip will be the easiest part to loosen. Because cornice molding is thin and long, it will bend out in the center without splitting.

Try to loosen each nail, removing with claw-hammer or pry bar. If the nail will not come out through the molding, keep pulling with the pry bar until the cornice pulls out from wall and ceiling. Casing nails are long and need a good deal of working to remove. When the first nail comes loose, work along the molding to the nearest nail in each direction, pulling each out a little farther. Soon the middle nail will come entirely free.

By then the piece should come loose. At the last nail or so you can usually pull out the cornice strip by hand.

To replace, miter the ends of the new cornice piece and nail in with finish nails or casing nails. Countersink. Cover with putty, allow to dry, and sand for painting.

CHAPTER EIGHT

Flooring

The actual laying of a complicated, large-room floor should be done by a professional carpenter, to avoid various engineering hazards of load and stress factors. However, the amateur carpenter should know the theory of floor-laying in order to make repairs of his own.

It is, moreover, quite appropriate for the home-owner to lay a new floor *over* an old floor, or build a new floor over an area in the basement or cellar that does not yet have one.

STRUCTURE OF A FLOOR

Essentially, the make-up of a floor does not differ much from the basic construction of a wall. It is composed of a skeletal framework and a covering. In the case of the floor, the covering is attached to the top of the framework, rather than to the side.

Instead of studs, which hold a wall up, a floor is supported by long framing members called **joists.** A floor joist is usually dimension timber 2 inches thick and 6, 8, 10, or 12 inches in width. The width depends on the expanse of the floor and the weight of the surface material.

The joists supporting the floor of a frame house usually extend from two sides of the concrete foundation, the load-bearing walls. If the width of the house demands it, the joists may be spliced together over a center foundation or series of concrete pilings. Or they can be spliced together over a girder running parallel to the two load-bearing foundations. The girder is usually a piece of dimension timber twice as thick as the joists and as wide or wider. Joists 2 by 10 would probably be supported by a girder 4 by 12, for example. The 4 by 12 girder might well be two 2 by 12s set on end and nailed together.

In order to keep floor joists directly upright, 2 by 4s called **bridging blocks** are nailed in place between parallel joists forming an X. One or more Xs are used per joist at intervals of 10 or 12 feet.

Attached to the joists is a surface of **subflooring.** This layer is actually underlay material that supports the floor surface—be it wood, ceramic tile, vinyl tile, linoleum, rubber tile, or carpeting.

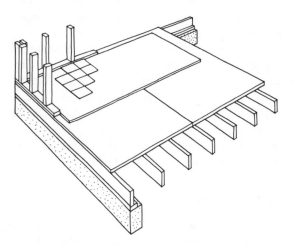

Fig. 111. Typical floor in a wood-frame house is composed of subfloor, underlayment, and a flooring surface as shown in the drawing. Subfloor can be made of shiplap strips or plywood, with underlayment of plywood or particle board. Flooring surface can be tile, carpet, linoleum, or other nonstructural flooring.

The two basic types of subflooring are **board strips** and **plywood.**

Board strips used for subflooring are usually either **shiplap, tongue-and-groove,** or **square-edge.** Since they serve mainly as a fastening surface, they can be Number 3- or Number 4-grade lumber. Subflooring is laid across the joists either at right angles, or diagonally to give rigidity to the finish flooring and to allow the finish flooring to run either parallel or at right angles to the joists.

On top of the subflooring a type of plywood secondary subflooring called **underlayment** is occasionally attached. It gives structural rigidity to the flooring and creates a solid base for heavier finish floors or for non-wood materials.

A layer of building paper is placed between the finish floor and the subflooring or underlayment to act as a cushion for the finish flooring, as a sound block to keep the two layers from rubbing against each other, and as a moisture barrier to keep spilled water from seeping through into the subflooring.

The finish flooring is then placed over the building paper. In some cases, like vinyl flooring or linoleum, the finish surfacing is attached with adhesive directly to the underlayment for a smooth bond. Indoor-outdoor carpeting can also be attached directly to underlayment and bonded with adhesive. Floor strips are attached by nails to subflooring or underlayment through the building paper in the conventional manner.

The floors of second stories are constructed in the same fashion, with the joists spanning the width of the house, sometimes spliced together over a middle bearing wall. The ends are secured to the exterior load-bearing walls of the house. Second-story joists not only support the subflooring and flooring of the top story, but also act as a nailing surface for the ceiling below.

SUBFLOORING

Subflooring strips can be laid either at right angles to floor joists, or diagonally to them. Right-angle subflooring will support only board flooring running parallel to floor joists. Diagonal subflooring, however, will support board flooring running either parallel or at right angles to floor joists. This type of application requires cutting each board at 45° angles rather than straight across.

Plywood subflooring can be laid either parallel to or at right angles to joists. Although strip subflooring can be used to underlay tile and linoleum, it is advisable to use plywood subflooring to support the several types of non-wood flooring surfaces: vinyl tile, linoleum, rubber tile, ceramic tile, and carpeting.

Strip subflooring is face-nailed into position, using two 10d nails to a joist. Start right-angle flooring strips at one wall, making sure that if more than one strip is used in a run, it will meet a joist on-center. A break in the adjoining run should meet at another joist. All subflooring strips must meet at joists except for end-joined tongue-and-groove stripping, which can join anywhere.

Start the subfloor at the sill plate and end it at the opposite sill. The last strip laid may have to be ripsawed to fit.

In the case of joists that are not entirely true, it is best to plane off the portions that stick up and to shim up the sags before applying subflooring. Unevenness in subflooring tends to make the flooring itself all the more subject to squeaking later on when the wood settles into place.

With plywood subflooring, start in one corner with a module, securing the plywood to the joists with nails 8 inches apart. Cut each succeeding module to fit as necessary.

FLOORING

Once the subfloor is in place, lay 30-pound felt building paper over the entire surface before laying the flooring strips. Be sure each piece of paper overlaps several inches. Building paper keeps the two layers of wood from rubbing against one another and producing groans and howls when the floor is walked on. Paper is simply laid down on the floor—no special installation is necessary.

Wood flooring comes in various styles and types. It can be **hardwood** or **softwood.** Oak, beech, and maple are usually considered hardwood floors, although some pine and fir, which are softwood, are harder than some maple or beech floors. Oak is the most commonly used and by many is considered the best of woods for strip flooring.

Wood strips come in various widths, depending

on the type of wood and the particular kind of lumber cut. You can get square-edged floor strips, called **planks, tongue-and-groove,** and **block flooring.**

Planks are wide board strips designed to imitate the type of flooring used in old Colonial houses. The strips are supposedly fastened to the studs with wood dowels. To make the planking look real, false round dowels are bonded into the plank strips. Ordinary flooring nails are used to fasten the stripping down. The planks are not jointed together, but have square ends and sides.

Tongue-and-groove floor strips are fastened together in a tight fit that keeps the strips aligned and prevents bulging or warping of an individual strip. The tight joint of tongue-and-groove also tends to keep the floor tight against moisture and sound.

Block flooring consists of squares or rectangles that when laid resemble parquet surfacing. Because of the design of the squares or rectangles, the floor has a much different appearance than a conventional wood strip floor.

Tongue-and-groove and square-edged flooring strips can be applied to any kind of subflooring, but block flooring should be applied to plywood subflooring only.

Like any type of wood used in home carpentry, board flooring should be allowed to acclimate itself to the room in which it is to be laid. Have the material delivered to the house several days before you are going to lay it. Stack it off the subfloor on 2 by 4 props and let it season slightly in the usual temperature and moisture conditions to which the room is subject.

If the room is small enough, 10 feet by 12 feet or so, you can begin to lay the flooring strips or planks against one wall, and work across. If the room is a large one, it is best to begin in the center and work both ways out.

NAILING FLOORING STRIPS

Nailing varies with different types of wood flooring. For tongue-and-groove strips, blind-nail with one nail every 10 to 15 inches.

Nail sizes depend on the thickness of the flooring. Use 8d nails for $^{25}\!/_{32}$-inch flooring, 6d for $\frac{1}{2}$-inch, and 4d for $\frac{3}{8}$-inch. Use only flooring

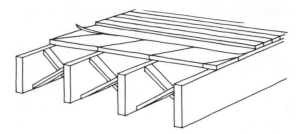

Fig. 112. Flooring of board strips can be shiplap, tongue-and-groove, or square-edge. Finish flooring can be attached directly to subflooring, with building paper between. Note bridging blocks beneath subflooring to keep parallel joists upright.

nails or nails with ringed or annular grooves that will hold the strips tightly in place.

To **blind-nail,** angle the flooring nail at 45 degrees through the point where the tongue joins the strip. When the next groove fits over the tongue, the nail head will be hidden—thus the strip is blind-nailed (see Fig. 95).

As a rule of thumb, nails should penetrate through the subfloor into the joists a distance twice the thickness of the subfloor, especially where plywood subflooring is used.

In general construction, the nails should always be driven into joists. However, if the subflooring is already laid, and you are putting down a new floor, you may not be able to locate the joists. If you have access to the area directly below the subflooring, you can measure from the wall and locate the joists in that manner.

If you do not have access to the area directly under the floor, simply nail directly into the subflooring. By staggering the position of the nails, you will achieve a satisfactory fastening pattern.

LAYING TONGUE-AND-GROOVE FLOORING

If you start at one wall, lay the first strip ⅜ inch from the edge of the sill to provide for swelling due to climate and seasoning or warping. Make sure the piece is parallel to the wall and at right angles to the adjoining walls. When the first piece is in position, face-nail it with 8d finish nails to the studs, using two nails for each joist.

Countersink the nails and fill the holes with plastic wood.

If you start in the center of the room, face-nail the first strip in place, leaving ⅜ inch at each end. Lay the groove-end along the center, and prepare a narrow strip of wood called a spline to fit into the groove not only of that strip but of its neighbor, which will be laid the opposite way (Fig. 114). The tongue of the average piece of 1-inch flooring tongue-and-groove strip flooring is ¼ inch in width. So are the overlaps. You can make a spline out of a piece of regular ¼-inch plywood by sawing it from a long piece.

Once the first piece is laid, push the second against it as close as possible and blind-nail it against the first piece.

LAYING PLANK FLOORING

For square-edged flooring, start at one wall, leaving ⅜-inch breathing room for the first piece. Face-nail each strip every 8 to 10 inches along its length. On 2-inch-wide flooring strips, use two nails no closer than ½ inch from each edge. For 4-inch-wide flooring, use four nails, and for 6 inch, six, and so on. When nailing planks, always countersink the heads and fill the holes with plastic wood.

After laying six planks, press a piece of scrap lumber against the last laid strip and hammer hard to press the pieces together tightly. Be sure all lengths stay in alignment as you do so. Repeat this every half-dozen runs.

Pause just before you get 2 feet from the end of the room. Measure carefully at each end of the run to make sure the last pieces are going to approach the opposite wall exactly parallel to it. If the strips seem to be veering in one direction or the other, plane off a bit at one end of the next strips so that the last piece parallels the wall. On tongue-and-groove, plane only the groove end.

Stop within ⅜ inch of the wall. The ⅜-inch gap around the room will allow for shifting and breathing of the wood. The gap will be covered by baseboard molding.

LAYING BLOCK FLOORING

Block flooring is especially easy to lay. Leave ⅜ inch around the border, and begin in the cen-

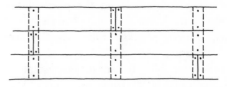

Fig. 113. Diagram shows typical installation of floorboards on joists. Note that all breaks occur over stud. Two nails are face-nailed into each board at break. When board runs over a joist, two nails are sufficient to hold it.

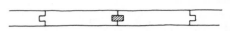

Fig. 114. Diagram shows how spline is inserted between two pieces of tongue-and-groove floor strips at middle of room. Spline for 1-inch boards—which are actually ¾ inch thick—can be made from narrow strip of ¼-inch plywood.

ter of the room. To determine exactly where to place each block, make a plan of the room on graph paper, using one square for a foot of floor area. By measuring the room you can find out how much space will be left over at the end of each run.

Face-nail block flooring to plywood-paneling subflooring so that it goes through and into the joists at least one inch. If you are applying block flooring to an extant floor or to a concrete surface, use adhesive and follow directions on the adhesive container. Block flooring is ideal for attaching to concrete surfacing.

To attach block flooring to a concrete slab, use an adhesive called mastic specifically formulated for the purpose, applying it to the entire surface of the concrete. Cover the surface with 30-pound felt building paper. Apply another surface of mastic to the building paper. Lay the block flooring directly on the second layer of mastic.

Parquet-type wood blocks are also available in 12-inch squares with ⅛-inch-thick layer of resilient foam on the underside. The tiles consist of four laminated plies or panels, each $\frac{5}{16}$ inch thick. In addition to the foam backing, each block has small islands of pressure-sensitive adhesive.

Fig. 115. Photograph shows plank floor. Planks have appearance of being held to floor joists by dowels drilled through flooring and into framing members. Planking dowels are actually supplied at mill and flooring is applied with ordinary flooring nails. Photo courtesy California Redwood Association.

Fig. 116. Parquet tiles resemble chessboard patterns when laid. Photograph shows a parquet-type design on a floor. Squares can be as small as 6″ or as large as 2′. Photo courtesy Armstrong Corp.

To install, you peel off the protective paper backing and press the squares in place; no nailing or spreading of adhesive is necessary. Installation is similar to that of pressure-sensitive vinyl tiles.

These tiles can also be laid over wood flooring and vinyl-tile flooring—any surface that is solid and clean.

LAYING WOOD STRIPS ON CONCRETE

While it is not possible to attach ordinary tongue-and-groove wood flooring strips directly onto concrete, a substitute type of flooring made of three-ply oak planking ⅜ inch thick can be applied directly to a concrete floor with special mastic adhesive.

These planks are made of oak facing in a chestnut-brown finish that includes factory-installed walnut pegs. The lengths come in assorted sizes, and three random widths of 3, 5, and 7 inches.

For best results, follow the directions on the container of adhesive.

LAYING RESILIENT FLOOR TILE

Laying non-wood floors is actually not within the province of the carpenter, but most homeowners can do the job easily.

Ceramic tile is a complicated type of floor, and should be laid by a professional tiler. Terrazzo is a special floor tile made up of small marble chips and rock scraps set in Portland cement and buffed flat. A professional should handle this type of flooring.

However, resilient tile floors are much easier to handle. These include asphalt tile, rubber tile, vinyl tile, and linoleum.

Asphalt tile is a mixture of asbestos fibers, lime rock, inert fillers, and colored pigments with an asphalt or resin binder. The tile is brittle, and is bonded to the floor with mastic, either directly or over a layer of 30-pound felt building paper.

Rubber tile is composed of rubber, asbestos fibers, or some type of plastic stiffener, and is colored by mineral pigments.

Vinyl tile is similar to asphalt tile, except that vinyl-type resins are used as a binder, instead of asphalt.

Linoleum is actually linseed oil in a solid state. ("Lin" for linseed; "oleum" for oil.)

All these resilient surfaces are basically plastic-type blends and are easy to work with. The beauty of resilient tile is that it can be laid on any truly flat surface and attached by an adhesive supplied by the manufacturer.

If the surface is a concrete slab, the tile can be applied directly to the surface by mastic. If the surface to be laid is an older floor, it is advisable to apply an underlay of plywood over the floor to take out any kinks or sags in the old surface. Use underlayment-grade plywood for the job. In some cases a good solid wood floor can be used for direct application. Be sure all bulges and sags are buffed down and smoothed over with filler material.

Resilient tile comes not only in squares but in larger segments, like 6-foot strips or wider. For the homeowner, it is best to select tile squares of 9, 12, or 15 inches for the easiest application. Laying them does not produce the problems that laying large sheets of resilient tiling does.

To start, be sure that the surface is smooth and flat. Remove all grease or oil stains. Find the exact center of the room and divide it into four equal parts.

Start work on one quarter section and lay one line of tile from the center of the room to one wall. When you come to the wall, you will find that the last space is smaller than a tile. Take a fresh tile, lay it over the empty spot, overlapping the last tile laid. Mark the edge at the end of the last tile, and cut the tile. Resilient tile can usually be cut with a utility knife or even scissors. Lay the last tile, leaving a slight breathing space (⅜ inch) between wall and tile. The baseboard will cover this gap.

Lay the second row and work line by line until the quarter sector is laid. When you come to the last row of tiles, cut each to fit, as explained, and finish up the quarter section. Then do the other three quarters in the same fashion.

In laying tile, apply adhesive along a row for one or two tiles at a time, and then lay the tiles on the adhesive carefully. Drying time will be noted on the container of the mastic. Be sure each tile is in correct position before drying time is up.

Certain types of resilient tile come in squares

Fig. 117. Typical resilient floor tile is made for either professional or for do-it-yourself installation, recommended for new homes or old. Colors are bright, and maintenance is minimum. Photo courtesy Armstrong Corp.

A

Fig. 118. Photographs show how to lay resilient self-sticking tile. (A) Divide the room evenly with chalk lines marked on the present floor. Snap chalk line as pictured. Next step is to peel the protective paper off the back of the tile (B). Starting at the center of the room, do one-quarter section of the floor at a time, working to the sides. Line up the tiles accurately (C) and press them into place one after the other. Cutting border tiles is shown in (D). Trim the tiles with a pair of ordinary household shears after laying them against the baseboard and marking them with a straightedge. Scene in (E) shows job completed. Photos courtesy Armstrong Corp.

B

C

D

E

with pressure-adhesive affixed to the back side, covered over with wax or polyvinyl paper. In installing this type of tile, no mastic is needed. Mark the center of the room as explained and start laying at the center, peeling off the paper and pressing the first tile down. Then proceed as above.

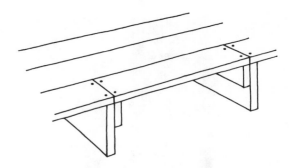

Fig. 119. Drawing shows how to replace a piece of worn-out or broken flooring. After removing the bad piece by sawing it up to the joists, fasten cleats to the joist to support the new piece. Then saw new piece to fit, and nail to cleats.

REPLACING DAMAGED FLOORBOARDS

When floorboards become damaged by gouges, rips, or other kinds of hard knocks, it is advisable to remove and replace the section of floor that is ruined. Select the particular board strips that must be removed and mark Xs on them. Then, choose one and bore a hole in the middle of the strip with a brace and bit, placing the hole as near the joist as possible. With a keyhole saw, a saber power saw, or a power trim saw, cut across the middle of the board, working with the grain. Pry up the two halves and remove the broken piece completely.

Remove as many more boards as is necessary.

Cut new boards to the sizes required by the opening in the floor. Square the ends with a saw, shaper, and plane. Trim off the good boards exactly at the end of each joist. Fasten a 2 by 4 cleat against the joist to support the end of the new board or boards. Fit the new boards in, face-nail them to the cleats, countersink the nail-heads, and fill the holes with plastic wood. Put back new wood strips and cover all the open holes as already indicated.

It is more difficult to remove a piece of tongue-and-groove floor stripping. Start the same way as with square-edge planking: drill a hole with a brace and bit in the center of the board to be removed. Then with a keyhole saw, power saber or power trim saw cut lengthwise with the board. Using a wrecking bar, pry the board apart and pull out each half. Fasten cleats to the joists and prepare for a new strip. To replace the tongue-and-groove piece, see page 95 in Chapter Seven explaining how to replace tongue-and-groove board paneling.

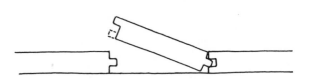

Fig. 120. Broken tongue-and-groove floor strip can be replaced easily by removing bottom side of groove before inserting tongue into groove. Press down groove side and face-nail to subflooring, sheathing, or stud.

SQUEAKING FLOORBOARDS

Another common problem that wooden floors develop is a squeaking board. The problem is either that a nail has pulled out of the joist or the joist has warped away from the floorboard and has taken the loosened nail with it.

There are several remedies. If the floor is made of planks, remove the squeaking plank and investigate the joist to which it is attached near the squeaking. If the joist has warped downward or sideways, fashion a "shim," or narrow strip of wood fill like a shingle or thin ⅛-inch piece, and fasten it to the joist so that the plank when replaced will adhere tightly to the shim at its proper level. This should stop any noise.

If the joist is properly in place, and if the board persists in squeaking, drill a screw hole in the board at the point near the joist close to the squeak. Use a brace and bit size just larger than the screwhead, and cut out a shallow well just above the screw hole. Insert a wood screw, tightening it until the floorboard is clinched to the joist. Fill the well above the screw hole with plastic wood, allow it to dry, and sand it down. Then finish off the wood to match the rest of the floor.

If the squeak is minor, simply face-nail several annular-ring finish nails, countersink, and add plastic wood. Sand and finish to match.

CREAKING STAIRS

Each step of a stair is made up of a horizontal board called a tread and vertical board at the back called a riser (Fig. 122). The inner ends of treads and risers rest on two notched timbers running diagonally from the bottom to the top of the stairs.

The treads and risers are secured by wedges glued into the grooves. As the stairs age with usage, the wedges loosen and the treads spring away from the risers. As you walk along the treads, you force them down onto the risers and the ends move in the grooves, causing squeaking and groaning.

The way to cure a creaking step is to secure the tread tightly to the riser that supports it. Drive 2-inch cement-coated finishing nails solidly through the tread into the riser. Drive these nails in pairs at opposite angles, toenailing them, then sink the heads below the surface with a nailset and press in plastic wood. Sand when dry.

If the nails don't work, countersink a hole in the tread, drill a pilot hole, and screw in a long, slender, flathead wood screw to secure the tread permanently. Fill the hole above the screwhead with plastic wood. Allow it to dry, and sand for finishing.

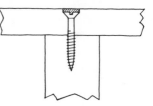

Fig. 121. Picture shows how wood screw can fasten loose flooring strip to joist below. Screw head is countersunk below level of floor surface, with space filled with plastic wood.

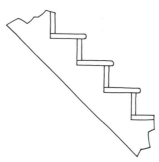

Fig. 122. Diagram shows how treads and risers are fitted together in inside stairway. The framing member that runs diagonally from the top floor to the bottom of the stair is called the stringer.

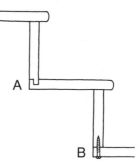

Fig. 123. Schematic drawing shows two different ways to connect tread and riser. (A) The vertical tread is rabbeted at each end, and the two risers are mortised into the treads for a tight joint. This fit minimizes squeaks and groans. (B) Tread is simply screwed to bottom riser in plain end-butt joint. If screw is loosened by constant wear, it must be replaced with a longer size. Some treads and risers are simply nailed together.

CHAPTER NINE

Doors and Windows

If the doors and windows in your house are warped, rotted, loose, or in any way in imperfect working order, they will constantly annoy you by their inability to function. It is one of the home-owner's primary duties to keep them in operation, making sure that the locks and latches are striking properly and that they slide, swing, and shut properly at all times.

ANATOMY OF A DOOR

There are three basic components to a door: the door itself; its hardware, including latch and hinges; and the frame in which it is hung.

Although there are many different types of doors, most of them are constructed along the same pattern, and almost all have the same parts.

Vertical pieces running the length of the door are called **stiles.** Horizontal pieces running across the bottom, the top, and through the middle are called **rails. Panels** are of wood, glass, or screen. The joints between the panels are sealed by **molding.**

The currently popular flush door is actually not a solid door at all, although it appears to be. It is made up of a loose frame of two stiles, top and bottom rails, with hollow spaces in between, and with each face covered with strong plywood. Both front (latch side) stile and back (hinge side) stile are wide enough to accommodate the proper hardware.

In the average door there are hinges, latches, and knobs, and in many doors either sliplocks or keylocks. There are either two or three hinges to each door, depending on its position in the house.

Fig. 124. Diagram of typical door shows all parts: stiles, rails, and panels. A flush door contains the same essential components, although they are hidden by the outer veneers.

Fig. 125. Schematic of door in its framework is shown from the top. Casing material—jambs and surface members—that covers the inner structural frame is millwork, but studs are framing material. Stop keeps door from swinging through frame. Jambs and stops run across top of door, too.

Outer doors contain three hinges, and inner doors two. A dutch door, of course, has four; two for each half.

The doorframe is the wall construction that surrounds the door. It consists of the **rough frame** and the **finish frame.** The **trim** or **casing** which you see when you look at a door is a finish border that covers the joint between the rough frame and the wall surface ending at the frame.

Jambs on the inner face of the frame are casing members attached to the rough frame. It is to the jambs that the hinges and the strike plate for the latch are attached. **Stops** are molding pieces against which the door closes; these are attached to the top and side jambs. The **sill** is the bottom tread of the doorframe.

HOW TO HANG A DOOR

Although it is a simple matter to take down a door, hanging one is a bit more complicated. There are three basic steps to hanging a door. The first step is the shaping of the door to the proper size to fit the frame. The second is the attachment of the hinges. The third is the installation of the lock latch to the door and the strike plate to the jamb.

To shape a new door to fit a frame, place it first of all with its back stile resting lightly against the doorjamb where the hinges will be attached. Mark a line along the door stile if it passes the opposite jamb to show how much material you have to remove to make the door fit. Plan to leave a $\frac{3}{16}$-inch gap all around the door to allow for opening and closing and changes due to weather.

Plane both vertical stiles down to the proper width. Then stand the door between the jambs and mark a line along the top, allowing the $\frac{3}{16}$-inch gap at both top and bottom. Plane or cut off the excess. Rework until the door fits easily into the frame against the stops.

To attach hinges, position them about 10 to 12 inches from the top of the door and about 12 inches from the bottom along the back stile. Mark an outline of each hinge leaf to indicate the borders of the cut. Then, on the side of the door stile, mark the thickness or depth of the leaf.

With chisel and hammer, make the undercut along the marked lines, and pare out wood segments carefully, as described on page 18 in Chapter Two. Fit the hinge leaf to make sure it is flush with both edge and side.

Then mark screw holes and attach both hinges with the screws provided, adjusting them until they are perfectly flush. When both top and bottom hinges are attached to the door, join the attached leaves to their mates, insert the pins, and place the door in the frame against the stops. Slip

a thin shingle or small magazine under the door at the bottom to take up the $\frac{3}{16}$-inch gap that will be there when the door is hung.

Holding the door in open position to the frame, place the leaves of the hinges against the jamb. Mark the hinge positions. Remove the door, separate the leaves, and outline each leaf carefully on the doorjamb to indicate the cutout.

With chisel and hammer, remove the excess wood. Drill screw holes and secure the hinge leaves with screws to the jamb. Attach the top leaf of the door to the top leaf of the jamb, insert the hinge pin, and see if the bottom hinge leaves will meet correctly. If not, adjust the bottom leaf. Then insert the bottom pin.

HOW TO INSTALL A LOCK

To attach a mortise lock (Fig. 126), which is usually the type used on an outer door, hold the lock against the side of the front stile and mark in the knob spindle hole, the keyhole, and the lock edge. You will find instructions with the hardware. Mark the borders of the lock cover plate against the door's edge.

With a brace and bit, drill holes of the correct size for both the keyhole and the knob spindle through the stile of the door. At the stile's front face, drill several starting holes and chisel out a rectangle to receive the lock part.

Install the lock, securing it in place with screws. Attach the keyhole, doorknobs, and spindle plate to the door. Close the door, and operate the lock and latch.

Mark the latch plate on the jamb at the proper place. Outline the borders of the plate against the jamb, using the latch plate itself as a guide. Chisel the marked segment out for mortising. Test for the right depth by working the latch and lock in the wood before attaching the plate.

When the position is verified, attach the strike plate to the jamb, making sure it will hold the door tightly against the stop when it latches.

To **attach a bore-in cylindrical lock,** designed for use with interior doors, bore holes as specified in the instructions that come with the lock. The lock will probably be provided with a template, which is a guide drawing showing you exactly where to drill the holes. Mark the door with the template. Drill both holes as shown in the stile

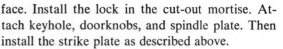

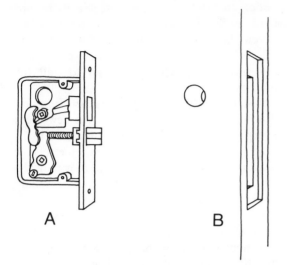

A B

Fig. 126. Typical mortise lock comes in two main pieces: the lock assembly shown in (A), and knobs and spindle. (B) Shows how mortise cuts should be made in door to accommodate lock section. Template comes with hardware to show what size cuts to make and where to drill them in the door.

face. Install the lock in the cut-out mortise. Attach keyhole, doorknobs, and spindle plate. Then install the strike plate as described above.

To **attach a cylindrical or tubular lock,** position the template around the front edge of the door. Drill the holes where shown, install the latch bolt, and insert the doorknob spindle through the connection. Secure it by fastening on the opposite doorknob with the spindle screw.

Then install the strike plate as described above.

ADJUSTING A DOOR

Installation of a door is only the first step. Afterward, it may have to be adjusted because of changes due to seasoning. Even after adjustment, it may need to be readjusted as the hinges pull loose from the frame, as the strike plate slips, or as the door warps and twists at the stop to throw the strike plate out of position.

A sticking door is usually caused by swelling, by warping, or by some distortion in the shape of the door itself. Sticking can also be caused by a

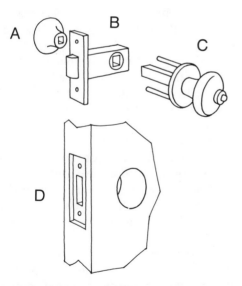

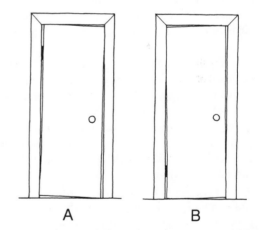

A B

Fig. 128. If door binds at the front top (A), tighten the screws of the upper hinge. If hinge is tight, remove lower hinge and shim it out with a thin strip of wood. If both hinges are tight, plane off the top front of the door until it fits snugly against jamb. If door binds at the front bottom (B), tighten the screws of the lower hinge. If lower hinge is tight, remove upper hinge and shim it out with a wood strip. If both hinges are tight, plane off the bottom front of the door until it fits snugly against the jamb.

Fig. 127. Tubular lock comes in separate parts: (A) outer knob; (B) face plate; and (C) inside knob. Latch plate for frame is not included. (D) Mortise cuts that must be made in door edge to take face plate. After plate is inserted, push inside knob through slot and secure in place. Attach outer knob. Template comes with hardware to show what size holes to drill and where to place them.

change in the hinge alignment, due to the settling of the wall and the frame on which the door is hung. The alternative to a sticking door is a door that rattles whenever there is a breeze or whenever someone walks across the room.

Both problems call for a removal of the door and realignment of hinges or strike plate.

To remove a door, open it and prop up the outer corner at the bottom with something like a thin book, magazine, or piece of shingling. By propping up the end, you take the weight off the hinges. Next, remove the bottom pin from the hinges, lifting it out from the top, or hammering it up with a nail or cold chisel laid against the underside of the pin's knob. Remove the top pin from the hinges in the same manner.

Lift off the door.

To put the door back, reverse the procedure, being sure to insert the top pin first. Then, remove the prop under the edge of the door, and put in the bottom hinge.

If the door latch in a sticking door strikes too closely to the frame, it is easier to reset the hinges than it is to plane down the front of the door and reseat the lock. To reset the hinges, simply remove them, chisel away the wood until each hinge lies deeper in the seating. Rehang, and make sure the door closes properly. If the door itself is too big for the frame, plane the back stile of the door.

If a door has shrunk so much that the latch does not catch at all, reverse the above process. Remove the hinges, and build up the seating with shims—strips of cardboard or thin wood—until the hinges are seated at the proper height. Be sure to use longer wood screws when you fasten the hinge leaves. The built-up cushion will demand more length in the fasteners.

A sticking door may be caused by the loosening of some of the screws holding the hinges. It is a good idea to tighten all the screws in all the hinges of your doors once or twice a year. The top hinges are under more strain, since they bear more weight by their position.

A rattling door may be caused by a latch that slides into the strike plate hole before the door is closed snugly against the stop. In a well-hung door, the latch should not close in the catch hole until the door is pressing firmly against the stop molding. To remedy this situation, don't remove the door. Just shift the strike plate slightly toward the side of the door containing the stop molding. Unscrew the strike plate from the frame, removing it entirely. With a sharp chisel, dig the hole out toward the direction you want to shift the strike plate.

Refasten the strike plate in its new position with wood screws in fresh holes. If the new holes are very near to the old ones, you may have to fill the old screw holes with wood pegs, glue, plastic wood, or even soft solder. Test the door until the latch does not catch before the door is solidly pressing the stop molding.

Sometimes the latch will not enter the strike plate at all, causing the door to fail to latch shut. Check first to be sure the door is hanging correctly. If it is not, readjust the hinges. If the door is hanging correctly, remove the strike plate and reset it, moving it away from the stop.

If a door binds at the top between stile and jamb, or if it scrapes on the floor, the top hinge, which supports most of the weight of the door, may have become loosened in the doorframe. Tighten the screws in the top hinges to pull the front stile away from the jamb. If that does not correct the problem, remove the hinge leaf in the jamb and cut away some wood. Refasten the hinge leaf.

If the door binds at the bottom between stile and jamb, tighten the screws in the bottom hinges to pull the front stile away from the jamb. If that does not work, remove the hinge leaf and cut away some wood as above.

If the door hangs too far away from the jamb at the top, shim out the top hinge leaf by inserting a thin strip of wood or several layers of cardboard before screwing in the hinge leaf with longer screws. If the door is too far away from the jamb at the bottom, shim up the bottom hinge recess.

If the doorknob slips and/or sags, the knob pin may be bent or the spring may be broken. Under constant use, the entire doorknob may become clogged with dust and rust. Use powdered graphite, not oil, to make it turn more easily.

To remove a doorknob to examine or lubricate, unscrew the knob from its stem and remove it. Take the stem out from the opposite end. If the spring is broken, replace it. If the trouble is caused by dust and dirt, clean it out.

When resetting hinges, it may be impossible to replace the screws with longer ones. Carve wooden pegs with a utility knife so that they are slightly larger than the old screw holes. Use the same kind of wood that is in the door. Smear the peg and screw hole with wood glue. Hammer the peg into the hole until it is snug. Allow the glue to dry thoroughly. Trim the ends of the plugs with a utility knife until flush. Insert the new screw.

Wooden doors may creak on their hinges if the hinges aren't oiled properly. Creaking can also be caused by the aging of the wood in the door. As wood ages, it tends to sag and warp. Twisted out of line, a door will not hang properly on its hinges. Because of that, the hinges begin squealing. Reset them as described above. Lubricate the hinges with ordinary oil and swing the door open and shut to distribute the lubricant through the hinges and pins.

ANATOMY OF A WINDOW

The two main elements of any window are the frame and the sash. The frame is built in the wall of the house; the sash is the wooden bordering that holds the glass panes. The sash either opens out or slides up and down in the frame when the window is opened or shut.

The principal parts of the frame which surrounds the window in the wall are the **sill,** or bottom piece; the **jambs,** or side pieces; and the **lintel,** or upper piece.

Supporting sill, jambs, and lintel inside the wall are horizontal **double headers** at the top and bottom, and vertical **double studs** at the sides. These pieces of framing timber are usually 2 by 4s nailed together side by side. The double headers and double studs give extra-strong support to the window frame.

The finish frame of a window is composed of the finish sill running along the bottom, the side jambs, and the head jamb along the top. In a double-hung window, the jambs are in turn composed of **stops,** or interior and exterior trim, and **parting strips.** The stop or trim is the edge against which the sash slides; the parting strip is the dividing strip between the two sashes. The head jamb has stops or trim and parting strip, too. The

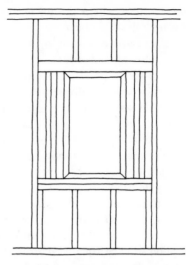

Fig. 129. Illustration shows the frame of a window as it is built into a wall. Double headers at bottom and sides are placed parallel to the studs across the width of the wall. The double header at the top that serves as the lintel is turned so that its strength is up and down, rather than across, to protect the top of the window from weight above.

Fig. 130. Diagram of the window sash shows all main parts: rails, stiles, muntins, bars, panes, sill, stool, apron, casing, and so on. Although the drawing shows a double-hung window, all windows are constructed with all or most of the parts shown.

sill includes a stool where the lower sash slides down for a tight fit at the bottom.

The principal parts of a sash, or window itself, are **rails,** the horizontal border strips; and **stiles,** the vertical border strips. The rails on the opposite ends of each window are called the meeting rails. A stile that separates the panes in the middle is called **muntin.**

There are four reasons why a window sash sticks in the frame and cannot be opened or closed. It may be jammed because it is paint-frozen; it may be binding because of swelling due to moisture; it may be sticking because the stop, or interior trim, is too close to the sash; or the sash itself may be too big either through moisture-absorption or excess paint layers.

To unstick a paint-frozen sash, first run the point of a utility knife between the sash and the stops all the way up and down both sides. Scrape away all extra paint. This action may unfreeze the sash. If not, remove the stop, clean it with a chisel or knife, and replace it.

To fix a sash that has swelled in the frame, locate the point where it binds, and force graphite or liquid window lubricant into the joint between the sash and the stop. If the sash will still not move, remove it. Plane down the area of the sash where it binds.

To unstick a sash pinned by a too-close stop, pry off the stop and renail it ⅛ to ¼ inch away from the sash. Be sure the gap is not more than ⅛ inch from the bottom of the sash when the sash once again closes to the stop, or weather-tightness will be lost.

IMPLEMENTS FOR GLASSWORK

Working with a window demands familiarity with material and tools used exclusively with glasswork.

The fasteners that hold a pane of glass against the sash or muntin are called **glazier's points.** The glazier's point is a small triangular piece of metal with a sharp point. You use a screwdriver to press it into the wood against the glass surface in order to hold the glass tightly in place.

Putty is a pliable, weatherproof material that tends to retain its doughlike consistency even after being in place a long time. It is used to seat a pane of glass against the sash. Putty not only seals out water, which might seep in between glass pane and sash, but tends to minimize shock to the window sash and protect the glass.

Putty is used frequently in other portions of the home, particularly in filling holes in wood before painting. It is formulated to take any kind of paint finish smoothly.

Made out of whiting and boiled linseed oil, it can be thinned by adding linseed oil. Pressure of the fingers can make putty quite pliable and ready to be applied anywhere.

Putty is applied with a special tool called a **putty knife.** It is a flat-ended pliant tool shaped like a kitchen knife, but with more spring to it.

Before applying putty, it is advisable in new wood to apply a quality priming paint to keep the linseed oil in the putty from flowing into the wood, leaving the putty dry and brittle.

To replace putty, remove all loose or crumbly pieces. Prime the wood as described. Apply the putty by making a long, thin roll and pressing it into place against the glass. Smooth it firmly with a putty knife.

HOW TO REPLACE A BROKEN PANE OF GLASS

Any number of accidents can break a pane of glass in a window. Replacing a glass pane is one of the homeowner's most common chores. Understanding the structure of a window sash makes the job that much easier.

The glass pane is held in the wood frame of the sash on one side by a "rabbet" cut, an L-shaped seat in which the glass is fitted, and on the opposite side by glazier's points, to hold it tightly to the sash (Fig. 131).

Usually the rabbet joint in which the glass pane is seated is on the interior side of the window, and the putty and glazier's points are on the exterior side. For this reason, you will do most of your work on glass panes outside, unless you can remove the window and work on it inside the workshop.

To remove a broken pane of glass, first scrape out all the putty holding the glass against the sash. Then take out the glazier's points with an old screwdriver. Wearing gloves to protect your hands, remove what's left of the broken glass.

Scrape the rabbet joint clean of putty, dirt, and

glass fragments. Measure the area to be covered by the glass pane. Deduct $\frac{1}{16}$ inch from the measurement to allow for wood expansion.

Purchase a piece of glass cut to the proper measurement. Store-cut glass will be more accurate than home-cut. Besides, it is a difficult job to do with ordinary carpentry tools and no experience.

Apply linseed oil to the rabbet joint of the sash—top, bottom, and sides—to ready it for putty. Soften a mound of putty with linseed oil and spread a bed $\frac{1}{8}$ inch thick on the seat. Make the bed very even so that there are no gaps.

Turn the pane of glass sideways and sight along it. You'll see that the glass curves slightly. Set the pane in place in the sash with the concave side inward, against the rabbet joint, so that the pane bulges outward toward the exterior of the window. Press the pane firmly in place. With a screwdriver, insert glazier's points to hold the pane securely against the sash.

If the glass does not tighten to the sash when the points are inserted, remove the points, press the glass more firmly against the thin film of putty, and replace the points (see Fig. 131 for guidance).

Roll a small piece of putty in your hands until it is pencil-shaped. Lay the roll along the outside of the glass pane, covering the glazier's points. Continue to apply the roll over top, bottom, and sides of the frame.

When the putty is in place, press a putty knife on it, moving it along against the glass and the end of the rabbet seat to form a smooth diagonal layer (see Fig. 131). When the putty is smoothed out against both glass and wood, let it dry. Then paint the putty to keep it from drying out and chipping off.

TIPS ON WINDOW REPAIR

A rattling window may be caused by a loose pane of glass or by a loose sash in the frame. You can always get temporary relief from the noise by sticking a piece of folded cardboard, rubber, or wood in the sash.

If the window rattles because of a loose glass pane, remove the putty, and the glazier's points, reputty, replace the points, and then spread the putty carefully against the glass.

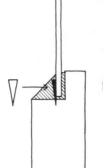

Fig. 131. Diagram shows the side view of a window with the pane of glass fitted in against the rabbet joint in the sash. The joint is the same along all stiles and rails. Note that there is putty on both sides of the glass, but glazier's points only on the outside.

In rare instances, the putty between the rabbet and the pane of glass may have dried out or fallen away. Remove the pane of glass, scrape out all the putty in the rabbet all the way around, and replace it, proceeding as described above in replacing a pane of glass.

WINDOW AND DOOR SCREENS

Because of heavy usage, window and door screens frequently tend to pull out of their wooden frames. A sprung screen not only lets in flies and other bugs, but is an eyesore. The amateur carpenter should be able to fix door and window screens.

First of all, remove the outer molding which covers the joint of the screen and frame. When the molding strip is removed you'll see that there is a tight groove in the frame into which the screen fits. Pull hard on the screening to take the wrinkles out of it and tuck the end back into the groove.

With a staple gun pushed hard against the screen, hold it in position and squeeze the trigger. Each staple will span several strands of wire and fasten the screen securely. When you've stapled all along the loose border, nail the molding back in place.

If the entire screen door or window has been

injured so that the screen is hopeless, buy new screening and start from scratch.

In order to fit the screen tautly into the slots for a smooth fit, it is necessary to bend the frame slightly while stapling either top or bottom into the grooves.

Lay the screen frame on a large table or workbench. Staple the screen at the top after working it into the slot until it is securely fastened.

Place strips of 1-inch wood under the top panel and the bottom panel of the frame. Apply large wood clamps to the middle of each stile of the frame, fastening them tightly to the surface of the table or workbench until the frame touches the table. The stiles are now curved in a bow shape.

With the screen bent, pull the screening tight and fit it into the bottom groove. Pull the screen as tight as possible, removing all waves and kinks. Then staple it in place.

Remove the wood clamps, and the stiles will straighten out, pulling the screen very taut in place.

Staple the sides and replace the molding strips all around.

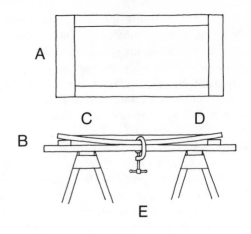

Fig. 132. New screen can be fitted into old frame by bowing frame and securing screen to ends. (A) Shows screen frame with old screen removed. (B) Shows how to bend screen frame into a bow by fastening wooden blocks under top and bottom and tightening the frame in the middle with C-clamps on each side. (C) Shows screen fastened at top of frame. (D) Shows screen fastened at bottom after being stretched tight across bow. When C-clamps are unscrewed, (E), frame will bend back to normal, pulling screen tight.

CHAPTER TEN

Storage Space and Extra Rooms

Storage space is always at a premium in any home, as is living space. In a consumer society, the average family today owns a great many more things than the family of fifty years ago. The problem with storage space, as well as living space, is knowing how to utilize it.

There are several types of storage space available in the home: closet space, drawer space, chests and bureaus, and shelf space. All these can be built by the amateur carpenter in order to provide not only more space to keep things, but to make the house neater and cleaner. He can also add a room in the attic, in the basement, or even in an unused portion of the house.

All he needs to know is how.

CLOSET SPACE

Although a closet is usually custom-built in a house to fill certain areas set aside by the architect for storage, you can build your own closet in a corner of a room or in a corridor between rooms. In a sense, it is robbing Peter to pay Paul, but sometimes the space set aside will be a life-saver for the housewife who has nowhere to stash her things out of season or when she isn't using them.

Fig. 133 shows how to construct a good, solid closet anywhere you have room.

Use 2 by 4 framing members 8 feet long to start the closet, nailing one sill to the floor and plates to the ceiling to outline the closet. Nail the vertical sill to the wall, and studs to the floor sill and plate. Toenail the 2 by 4 studs at the bottom and top to sill and plate, using 10d nails. Brace

the studs with a fire stop in the middle, and a nailer near the top to support a clothes pole if interior wall is gypsum board.

With the framework in place, attach panel strips to sill, fire stop, and plate, blind-nailing the pieces as described in Chapter Seven. You can cover the inside of the closet with thin cedar strips, if you like, or gypsum board for paint surface if you prefer it to match the other two walls. You can even leave the interior surface open if the outside paneling is presentable.

For the best results, install a prehung door in the doorframe. The unit can be purchased at a building-supply house. If you are adventurous, you can always build a frame and hang your own door, following the instructions on Page 110 in Chapter Nine.

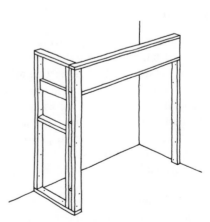

Fig. 133. Drawing shows schematic diagram of a simple closet in the corner of a room or open area.

If you choose to plan your own closet, note that the minimum inside depth of any closet should be at least 25 inches. The length depends on how much available space you have to work with. Most closets are room height—8 feet, or slightly less, depending on floor thicknesses and ceiling hangs.

ROOM DIVIDERS

A foolproof type of spacemaker is a room divider, which can be built easily by the amateur carpenter. It has the advantage of masking off an ugly part of a room, as well as giving privacy to an area needed for study or contemplation. One made of strips of wood is the simplest to build (Fig. 134). You can always design your own by a simple stretch of your imagination.

To begin the room divider pictured, nail a 2 by 4 to the floor and to the ceiling so that they are in direct vertical alignment with each other. Be sure that you have nailed each member at least 1 inch into joists in the floor and ceiling.

Using 1 by 4 pieces of lumber as shown, nail the first piece into one wall.

Then nail the blocking pieces—whatever length you want—into the first piece.

Nail a long piece and alternately assemble the divider until you have filled the area as shown.

Toenail in the last piece of lumber to hold the divider in place. Using a nailset, punch the nail out of sight, cover the hole with plastic wood, let dry, and sand down for finishing.

CABINET DOORS

In some situations, a storage space can be provided simply by building a door to cover an area where things are stored. This is particularly true of an open crawl space where a sloping ceiling prevents moving about close to the wall.

You can easily construct a door for a storage space or step-in closet by using strips of pine or cedar. Tongue-and-groove will give you the tightest fit, but planking will do almost as good a job.

To begin, assemble the pieces flat on the floor as shown in Fig. 135, with the tongues fitting snugly into the grooves. If you are using planks,

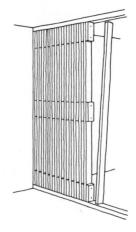

Fig. 134. The room divider shown is one of many that can be fashioned from pieces of wood and a bit of imagination. The text explains how to make the one pictured.

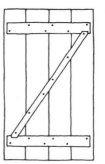

Fig. 135. The simplest kind of tongue-and-groove door can be made with a minimum of labor and materials. See directions in text.

fit the strips tightly together until the pieces are aligned perfectly. You can glue these pieces together if you wish.

Lay a 1 by 2 crosspiece cut 2 inches shorter than the width of the door and center it on the strips 3 inches from the top. Tack the crosspiece onto the strips. Repeat the same procedure at the bottom.

Place a 1 by 2 strip diagonally across the two pieces already in place, mark, and cut to fit. Lay the diagonal piece in place and tack it to the panels.

With the shaping boards accurately tacked in place, drill pilot holes for wood screws, countersink the holes, and screw the members together for permanent bond. If necessary, use a hand plane to trim off the exposed tongue-and-groove. Then hang the door with cabinet hardware—including hinges, latch, and knob.

DRAWER SPACE

A drawer is simply an open box or lidless chest that can be used to store household goods. The main point to keep in mind is the versatility of the drawer—in position, in size, and in installation.

A drawer should be placed so that you can stand almost directly in front of it and open it to its full depth. And you should be able to see into the bottom when it is open.

The size of the drawer is important in judging its usefulness. Here are some rules of thumb to follow when you begin to construct your own drawers: A drawer should be 12 inches high or less, 30 inches deep or less, should have handles near the sides, but not more than 3 feet apart. A heavy drawer should not be installed at the top of a cabinet without making sure the cabinet will remain upright when it is opened wide. A drawer should have stops installed so that it cannot be pulled out too far. All drawer slides with rollers should be equipped with stops.

Usually drawers are built into cabinets or chests. But they can be suspended under counters, tabletops, desk tops, and so on. Small drawers slide into racks screwed on top of shelves in kitchen cabinets.

Stacking drawers are available, too. These are individual wooden drawers, each in its own frame. The frames have sides, but neither top nor bottom. They are built so that they can be interlocked on top of one another. With these stack drawers you can build your own chest of almost any number of levels without using any nails or screws. You can even set them up side by side.

Every drawer needs some kind of framework to hold it. A typical drawer framework is made out of 1 by 2 strips nailed to vertical boards. For a typical installation, see Fig. 137. It is important that the sides be square for the easy sliding movement of the drawers. This framework is designed for drawers to fit flush with the guides. The back strip of the framework extends about ¼ inch above the side guides to serve as a stop.

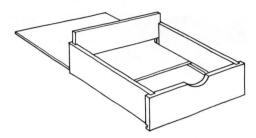

Fig. 136. Simple drawer contains a lot of joints, but the dado is the simplest to make. Note how drawer fits together after the cuts are made. It should be glued when finished.

Fig. 137. This drawer framework is made of simple materials and with a minimum of cutting.

This type of drawer framework can be attached to the side walls of a closet, or simply against a back wall. Or it can serve as an independent portable unit, which can be moved to any portion of the room or house.

A drawer is basically simple to build. First of all measure and mark the exact size you want—height, depth, and width. A dado is a groove cut into wood. The dado at the bottom of the front piece on the drawer illustrated in Fig. 136 should not be more than one fourth of the thickness of the strip. The same is true for the dadoes where sides join front and back pieces.

To make a dado joint, measure the depth of the joint and mark it on the wood stock used. Mark the lines where the cuts are to be made. Set a power saw's blade depth exactly as deep as the dado joint, and saw along the two cut lines.

Fig. 138. Photographs (A) and (B) show different ways to hang drawers in a framework. (A) Drawer slides into dado in frame and is held in place by a wood strip on side of drawer. (B) Bottom of the drawer slides into the dado in the frame, with stop provided by insert of framing member. Photos courtesy American Plywood Assoc.

Remove the remaining wood with a wood chisel and smooth the surface of the cut.

If you do not have a power saw, mark the piece in exactly the same way, and with a backsaw or crosscut saw cut carefully no farther than the end of the mark. Remove the waste with a wood chisel.

When fitting in the right-angle piece to a dado joint, use glue *and* nails for a tight permanent fit.

For the bottom of the drawer, you can use any kind of sheet material: plywood, hardboard, or thin wood.

CHESTS

A chest or bureau is essentially a portable piece. Occasionally chests and bureaus are built into the house itself, under a window seat, or under a bench, in which case the opening door should be in front—rather than a lid on the top.

Building a free-standing chest or bureau is a simple project. Hinge the lid far enough forward from the back of the chest or far enough in from either end so that when it is opened it will tilt away from the chest and you do not have to hold it up to keep it from banging down on your head or fingers.

In building any type of boxlike form, you can use rabbet joints at the corners where right angle pieces meet, or you can nail each member to a framing piece.

A rabbet joint is made exactly like a dado joint, as explained above, with one cut instead of two. After the depth of the rabbet cut is marked, set your power saw blade depth to the same measurement, and pass the saw through the cut. Remove the excess wood with a chisel.

If you do not have a power saw, make the same cut with a backsaw or crosscut saw, being sure to stop in time for a perfect cut. Remove the excess wood with a chisel and smooth.

HANGING SHELVES

Although at one time shelves were used only for holding books, today there are all kinds of shelves, and they are used to hold all kinds of things. There are fixed shelves that are attached to the wall, to the floor, or to a base of any kind. There are adjustable shelves. There are slanted

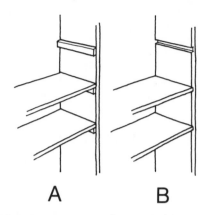

Fig. 139. Two different kinds of shelf mounts are shown: (A) by means of a strip holding up the shelf; (B) by means of a dado into which the shelf is slipped.

A   B

Fig. 140. Photograph shows how a large entry-hall closet can be split, allocating one side to family use and the other to guests. The left side contains over a dozen small drawers, shelf space, and pull-down shoe rack underneath built-ins. The right side can be maintained for guests with a full-length mirror, shelves, and garment brackets on doors which permit access to coats. Photo courtesy Western Wood Products Assoc.

shelves. There are drop shelves that fold down out of the way when not in use. There are revolving shelves like lazy-Susan kitchen cabinets. There are sliding shelves as in a refrigerator. There are lift-up shelves, like typewriter shelves used in office desks.

There are also shelves that are built into other storage units. There are shelves recessed into walls and shelves cantilevered from walls and posts. There are shelves suspended between facing walls, boards, or pairs of posts. There are shelves suspended between single posts, and shelves hung in wire or wood brackets.

When installing a shelf, there are three points to consider: The vertical members or uprights that support the shelf must be strong and steady; the horizontal shelf-hanger strips must be secure; and the shelf itself must be strong enough to bear the weight of the objects you expect it to hold.

If a shelf is to be suspended between two solid facing surfaces—as in a bookcase—you can nail the ends to wooden cleats fastened into the end supports. Or you can set them on adjustable L-shaped metal hangers that plug into holes drilled in the end supports.

For a temporary shelf, you can fasten two long, slotted, vertical metal frames to the surface of a wall, fastening the brackets that fit into the slots wherever you want. Wooden shelves can be laid on brackets with metal lips that hold them in place. Be sure that the frames are attached securely into a stud with long wood screws.

Shelves can be cantilevered from walls on a variety of plain or ornamental hanger brackets screwed to the walls in fixed positions.

You can even suspend shelves from the ceiling joists—usually in a basement or garage—resting them in U-shaped cradles of heavy wire or wood.

For a more artistic carpentry job, you can prepare wooden uprights with shelves running every 12 inches all the way to the top. Cut dado slots for each shelf to fit tightly into. Attach the uprights to the wall, and slide the shelves into place, fastening them at the slots.

Fig. 141. Plan is designed for all manner of shapes and sizes of items for storage on a wall catchall. Shelves are held up by strips of 1 by 2, with all members made of 1 by 12 pine, nailed with 4d box nails.

1-inch pine material of 8-, 10-, or 12-inch widths. The cleats pictured are made of 1 by 2 strips. To prevent the shelves from sagging, do not let them exceed 24 inches in length without full-width bracing.

A more complicated type of shelving construction is shown in Fig. 139B. The dado groove in the uprights should be exactly the thickness of the shelf material to allow the shelving to fit into the notch for perfect joints.

Bookshelves are usually 8 inches deep and 9 inches high for average-size books, or 12 inches deep and 12 inches high for larger volumes.

If you want a more versatile type of shelving for changing needs, you can buy special hardware for making adjustable bookshelves that can fit into notches an inch apart all the way along the uprights.

SIMPLE SHELF CONSTRUCTION

The simplest construction of shelving is shown in Fig. 139A. Uprights and shelves are made of

PEGBOARD

Pegboard is actually a kind of informal wall-storage space. It is hardboard either ⅛ or ¼ inch thick dotted with small holes at 1-inch intervals.

Mounted on a wall surface, the board becomes a perfect spot for hanging all kind of things on specially designed hooks, brackets, tool holders, small shelves, ties, and so on.

Each hook can hold considerable weight. A large board can hold hundreds of pounds of tools, utensils, implements, and so on.

Pegboard comes in panels up to 4 by 8 feet. Mount it ½ inch out from the wall so the hooks can be inserted. The hooks are cantilevered out, and grip the inside surface. You can use ½-inch furring strips to attach the pegboard to the wall if the surface presents problems for direct application. Nails holding pegboard should be 4d or 6d.

FINISHING OFF A ROOM

Many homes are sold with unfinished areas in the attic and cellar which are walled with bare studs or made of unfinished concrete blocks.

Finishing off a room is not nearly as complicated as attaching a separate room to a house. Finishing off is a big job, but it is a job that can be handled by the amateur carpenter without too much trouble.

If you want another room added onto your house—one that will necessitate a floor, outside walls and a roof—you should call in an architect or carpenter to do the job for you. Also, most towns require special permits for most of this sort of work. When the rough framing is in place— that is, the studs, sills, plate, subflooring, and roof —then you can finish it off yourself.

A room can be finished off in paneling strips, prefinished paneling, or in dry wall for a paint finish. In the following sections, two kinds of finishing are included, even though you may use only one of them when you go ahead to make your own room.

FINISHING OFF A BASEMENT ROOM

In many homes with large basements there is ample space for the construction of a large recreation area, playroom, or family rumpus room. Because bare concrete-block walls do not make for very attractive surroundings, it is a good idea to finish off the rudimentary structure with gypsum board or paneling.

Let's suppose the original cellar is occupied with furnace, hot-water heater, and other necessities at one end of its rectangular shape. Two long walls and a short wall are free, with a concrete floor over the whole area, and open floor joists above.

The idea might be to cover the concrete-block walls and build a new wall with a door in the middle across the end of it to keep the furnace and heater separate and out of sight.

Be sure when planning such a room to leave all utilities accessible for repair or replacement. It is not a good idea to build a wall too near a furnace or hot-water heater: The resultant heat and moisture can cause excessive warpage and even ruin paint jobs on gypsum board.

You should test out the cellar to make sure it is not a flood area during water runoffs. Moisture sometimes forms on concrete blocks due to condensation during certain weather situations. If such is the case, you should build with a special gypsum board containing a vapor barrier. The vapor barrier is a thin covering of reflective metal foil that keeps the moisture from seeping in through the gypsum.

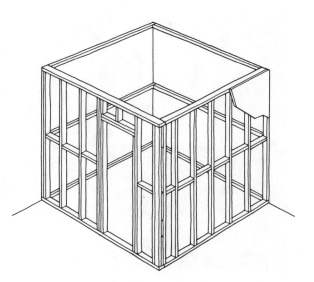

Fig. 142. Plan shows a typical finish-off room for the corner of a cellar or basement. In planning your own, simply move the door about where you want it, and follow the same instructions shown in the plan and described in the text.

Fig. 143. These photographs are before and after shots that might be your own attic space. Note how bay window and wall-to-wall window seat dominate one end of re-modeled attic, with the window seat containing storage space and room-size heating system. Paneling is 1 by 8 V-groove ponderosa pine, treated with custom stain. Photos courtesy Western Wood Products Assoc.

Most cellars foster dampness because of their position underground. If you plan to use the room throughout the year, you should bring heat into the area. The heat will help minimize evaporation moisture.

If the walls tend to sweat or leak, it is a good idea to give them a coat of waterproof epoxy, which you can get at any building-supply house. The epoxy will tend to cut down active leaks and to prevent a certain amount of natural seepage from the ground. However, if you have a large flow of water into the cellar area, think twice about trying to build there.

Assuming the cellar is normally usable, let's suppose you plan to finish off all four walls and ceiling with gypsum-board construction.

The first step is to design and build the wall that will enclose the room. Although carpenters usually build wood-frame walls on the ground and then stand them up, in a closed-in section like a cellar you can proceed by fastening the sill to the concrete floor, attaching a top plate across the floor joists, and then inserting the studs between them.

There is one complication in attaching a top plate to joists as described. If the cellar is rectangular, the joists probably cross the narrow part, wall to wall. Therefore, the joists run parallel rather than at right angles to the proposed top plate.

In that case, build the wall frame on the floor and then raise it into place, fastening the end studs into the concrete walls at both ends with masonry bolts.

Attaching sill, studs, and furring to masonry—concrete blocks and concrete—can be done in several ways. Furring is usually nailed to concrete blocks by means of masonry nails driven through pilot holes in the furring and attached to the concrete blocks.

Framing timber like 2 by 4s is attached by means of masonry bolts screwed into lead plugs. To insert the lead plugs, drill ½-inch holes in the concrete and the concrete blocks and insert the lead plugs in the holes. Drill pilot holes through the 2 by 4 material for the masonry bolts and screw the bolts through the framing timber and into the lead plugs. The plugs expand for a tight fit as the bolts enter.

Follow instructions in Chapter Seven on building an interior wall, and in Chapter Seven on putting in the door frame. When the studs are up and the frame for the door is in place, you'll have to get an electrician to hook up the light fixtures and convenience outlets. Flexible-steel electric cable is installed before any finish cover is applied to an interior wall.

Once the wiring is in, prepare the concrete block walls for gypsum-board application by nailing furring strips in the proper pattern. Since the gypsum board is put up horizontally, the furring strips should be fastened vertically with a central strip, and top and bottom strips firmly imbedded to attach the borders of the gypsum board to.

Once the walls are ready, attach the gypsum board as explained in Chapter Seven.

With the gypsum board on all the walls, go on to the ceiling. Apply gypsum board directly to the joists, nailing into each joist with dry-wall nails. Stagger the gypsum-board modules.

The material is heavy and clumsy to handle. If you have a helper, have him hold the gypsum board in place while you nail. If you are working alone, you can make two T-bars, each with a pole an inch longer than the distance from the floor to the joists. Use the T-bars to prop up the gypsum board at both ends while you work on it. With the props in place, nail the board to the joists.

With the walls and ceiling covered, lay the floor over the concrete surface. You can put down resilient tile, wood blocks, or wood strips directly on concrete surfacing. Use adhesive as described in Chapter Eight.

The door comes next. Hang the door as described in Chapter Nine. Then attach the hardware.

Prepare the surfaces for painting, and apply finish.

FINISHING OFF AN ATTIC ROOM

Many houses are built with attic space unfinished in order to provide a storage area or an additional room for the family as it grows.

These attic spaces are from 8–12 feet or more at the peak, with the roofs slanting down at a 45-degree angle or steeper to the floor at the walls. The rafters are usually uncovered, and any flooring is probably subfloor. The resulting triangular space is awkward to use for living quarters, but is ideal for stashing furniture and seasonal items when not in use.

However, such a space can provide an ideal area for one or more rooms. Each room, incidentally, must have at least one window for proper ventilation and light.

The first thing to do is to plan the amount of usable floor space for the room you want to close in out of the triangular space. It is obvious that the wall cannot be 8 feet high, or the room would be only a couple of feet wide. The most common method of overcoming this problem is to make 4- to 6-foot knee-walls along the low portions of the roof, leaving the triangular space beyond for crawl-space storage.

The room will have a 4- to 6-foot vertical wall, an expanse of ceiling running along the roof slope, and an 8-foot horizontal ceiling overhead.

The ceiling can actually run all along the roof rafters to the peak for a "cathedral" effect. It is a less-conventional design, but provides a feeling of spaciousness and freedom if that's what you're after.

Whether you plan a cathedral ceiling or a conventional 8-foot horizontal one, be sure to provide vents in the eaves of the attic in order to allow for free circulation of air. Otherwise you will leave your attic prey to a buildup of moisture during humid weather.

Whether or not you decide on a cathedral ceiling or a conventional one can depend on the slope of the roof. Some roofs have steeper slopes than 45 degrees. The thing to do is to take a folding rule with you and try to visualize the size of the room you want. Then pace off the width and length and hold the rule up to the ceiling to indicate the position of the walls and their height.

Then draw a plan to scale to get the exact measurements of the boards you'll want.

Once you've decided, draw a floor plan with marks for each stud, sill, and plate. You can use this in ordering wood from the lumberyard.

Let's assume you decide to build a room with an 8-foot dry-wall ceiling, 6-foot knee-walls of wood paneling up to the slope, and gypsum-board wall for the fourth side. The portion of the roof slope will be covered with gypsum board

too, and the extant wall of the house will be covered with wood paneling in keeping with the paneled knee-walls.

The first step is to lay the wall sills along the floor. If you're using vertical wood paneling, run studs up from the wall sills and attach each to a rafter on the side. Toenail each stud to the sill, and face-nail it to the rafter.

When the studs are in place, cut short pieces of 2 by 4 fire stops to act as cripple top plate between the studs. If the wood paneling is thin or apt to twist, attach short pieces of 2 by 4 fire stop halfway up the wall for nailing bases.

For the fourth wall, which must have a door, attach the sill to the flooring in a position where the studs can go directly upward and be face-nailed to the rafters there. See instructions in Chapter Eight for advice on the door space.

Cut ceiling rafters out of 2 by 4s to hold up the gypsum-board ceiling 8 feet from the floor. Attach these members to the rafters by face-nailing them from the side.

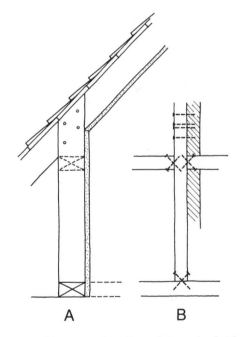

A   B

Fig. 144. Side view of wall stud in attic finish-off room (A) shows how stud is cut and fastened to ceiling joist from the side. (B) Shows same stud from front, facing wall, demonstrating how fire stops are inserted to provide nailing surface for dry-wall installation.

128   INSIDE WORK

The actual application of the paneling is simple. It is the preparation of the studs and sills that takes time and careful measurement.

The next step is to attach the paneling or the gypsum board.

The paneling should be attached to the 2 by 4 horizontals by blind-nailing through the tongues as described. For the outside wall, cut the panels to fit the angle of the roof, then blind-nail them in place.

Attach gypsum board to the ceiling first, to the roof rafters next, and to the fourth wall last, covering both sides identically.

Then tape and spackle all the gypsum board. See instructions on spackling in Chapter Seven.

Lay the finish floor on the wood subflooring (see Chapter Eight).

Attach baseboard–molding and wainscotting molding to the joint between knee-wall and roof, and cornice molding at the ceiling and roof (see Chapter Seven).

Hang the door as described in Chapter Nine.

Section IV

OUTSIDE WORK

A great deal of outside housework is too tough and technical for the amateur carpenter, but much of it can be handled. This section is divided up into two parts: exterior walls and roofs.

Although applying siding to a house is rightfully a job for an experienced carpenter, the knowledgeable amateur may be able to take it on with hope of success. The discussion in Chapter Eleven includes instructions on construction and on repairing siding of many different types, but not brick, stone, stucco, or metal.

Working on a roof is the province of the expert only, but the amateur can help locate a hidden leak, and he can also effect simple repairs like broken asphalt shingles, wooden shingles, and torn roofing paper. These are explained in Chapter Twelve.

Exterior Walls

Most homeowners have at one time or another seen a wood-frame house in the process of construction and have realized that it is formed of upright timbers that support ceiling and roof beams. However, it is difficult to visualize this skeletal framework after the wall surfacings are attached in place.

The anatomy of an interior wall is described in Chapter Seven. An exterior wall is constructed in the same manner as an interior wall, except that in a well-built house the inside of the wall is filled with insulation material or protected by insulation cover.

The outside surface of an exterior wall is quite different from an interior wall surface. Frequently, brick, stone, or stucco is used as a facing in front of the wall itself. This type of masonry construction is, of course, not in the province of the amateur carpenter. When the house is surfaced with wood, however—clapboard siding, shingles and shakes, and board-and-batten—then the exterior can be repaired or replaced by the amateur carpenter.

For the homeowner to repair a piece of wood siding accurately, he should know exactly how the wall is put together in the first place. And, in some serious situations of deterioration and ruin, it might pay him to put on a whole new surfacing over the old.

EXTERIOR SIDING

In spite of the apparent magnitude of the job, it is possible for the amateur carpenter to attach most new wooden siding materials to an existing framework or extant wall surface.

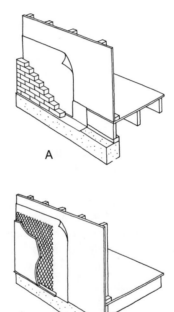

Fig. 145. Drawings show two typical exterior-wall surfaces, with details of brick veneer (A) and stucco veneer (B). Stone can be used as a veneer in much the same way. Note sheathing and building paper, separate studs, and outside layer of wall surface in each case.

Before beginning any kind of siding work, be sure to check the building codes in your area to see if a building permit is required. In most communities the owner of a house is permitted to work on any part of it, although anyone he hires must be licensed for the job. You may be putting your homeowner's insurance policy in jeopardy by hiring a moonlighting worker without a proper license.

Certain types of exterior sidings should be applied only to sheathing material. Other kinds do not need sheathing. In general, ordinary board siding and shingles and shakes require sheathing as a base, but sheet plywood siding does not.

The use of plywood siding can make a job easy since it goes on fast and provides an instant finished surface. The difference in cost of the plywood material will balance out the extra cost for labor.

Non-wood sidings like vinyl, hardboard, asphalt shingles, and aluminum generally require expertise that is beyond the ability of the amateur carpenter. It is best to write to the manufacturers of the products for information about application. They will tell you if it can be handled by the do-it-yourselfer or not.

One of the main requirements for the application of exterior siding is a good solid nailing base. If studs or the finish surface of a wall do not provide such a base, use furring strips either on the wall or on the sheathing. The patterns of the furring will vary according to the siding material and its application.

Fastening methods for siding are basically all alike. Furring strips should be attached with high-tensile strength rustproof nails long enough to penetrate at least 1 inch into the wall studs.

EXTERIOR SHEATHING

While new interior wall surfaces can be attached directly to the studding, some exterior siding materials do not have the proper strength or rigidity to stand on their own. To provide structural integrity and to help out with insulation factors, an underlay of sheathing is usually provided for an outside surface.

In houses already finished, the sheathing is there under the outer surface. For the amateur carpenter putting on a new siding cover over an old one, the old exterior itself can usually serve as a base for the new material.

If the old siding for any reason must be removed before being replaced, it is a good idea to know and recognize the types of sheathing that may be underneath it. The main types of sheath-

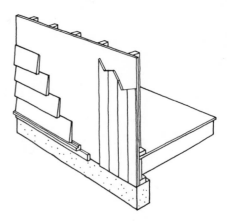

Fig. 146. Wood siding can be applied in either horizontal or vertical strips. Picture shows bevel wood in position; shingles are applied in the same manner. Vertical siding is usually tongue-and-groove or shiplap, although it can be of regular plank composition in board-and-batten or board-and-board applications.

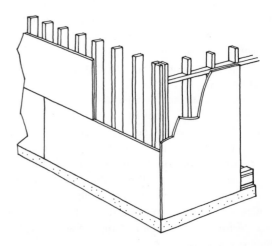

Fig. 147. Illustration shows how plywood sheathing can be applied to exterior of house both in horizontal and vertical positions. Fiberboard and gypsum board can also be applied in the same manner.

ing materials are plywood, exterior fiberboard, exterior gypsum board, and wood-strip sheathing.

The kind used most for structural situations to afford strength and protection is plywood. Its modular rectangular shape gives it ease of application and affords structural strength in both length and breadth. Diagonal wall bracing, com-

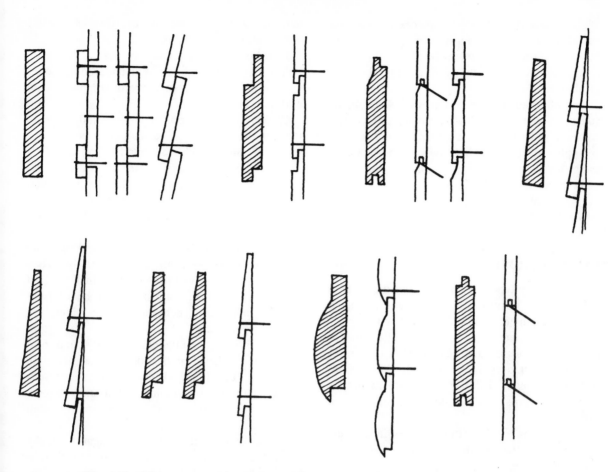

Fig. 148. Different types of exterior board siding are shown in the drawings, picturing the shape of the board in side view, along with smaller details showing nail installation.

mon in wood-frame construction, is not needed with plywood sheathing.

Plywood, fiberboard, gypsum board, and wood-strip sheathing should all be nailed directly onto the studs with waterproof spiral nails long enough to penetrate at least 1 inch.

Exterior plywood in square-edge or tongue-and-groove comes in $\frac{5}{16}$-, $\frac{3}{8}$-, and $\frac{1}{2}$-inch thicknesses, in panels 4 by 8, 9, or 10 feet.

Apply plywood as sheathing either vertically or horizontally. Nail every 6 inches along the edge and 12 inches apart to the center studs.

Use STANDARD grade exterior plywood. When applying, leave the ends spaced $\frac{1}{16}$-inch apart and edges $\frac{1}{8}$-inch apart.

Plywood offers excellent rigidity but low insulative value.

Exterior fiberboard in tongue-and-groove or shiplap patterns comes in $\frac{1}{2}$- to $\frac{25}{32}$-inch widths in 2- by 8-foot panels.

Apply fiberboard as sheathing horizontally with roofing nails 3 inches apart along the edges and 6 inches apart intermediately. Do not nail closer than $\frac{5}{8}$ inch from the edge.

Fiberboard offers fair rigidity and very good insulative value.

Tongue-and-groove exterior gypsum board comes in $\frac{1}{2}$-inch widths, and 2- by 8-foot panels.

Apply as sheathing vertically and horizontally with gypsum nails every 4 inches around the edges and every 8 inches intermediately. Gypsum board cannot be used as a nailing base for siding, but provides a good paint surface.

Gypsum board offers good rigidity, but low insulative value.

End-matched tongue-and-groove solid board comes in 1 by 6 widths. Shiplap solid board comes in 1- by 8- and 1- by 12-inch widths.

Apply as sheathing diagonally or horizontally, using 3 nails per stud for widths of 8 inches or more, 2 per stud for lesser widths.

Solid board provides good rigidity, and fair insulative value.

PAPER BARRIER

Between any type of sheathing and the exterior surfacing of a wood-frame house there is a layer of building paper, usually 30-pound building felt construction paper. The purpose of the paper is to act as a moisture and heat barrier, preventing moisture from seeping in through the exterior surface and being absorbed by the sheathing and transmitted across the wall space to the interior surface.

The paper is applied after the sheathing is up, and should overlap so that no point of sheathing is exposed.

SOLID-BOARD SIDING

Solid-board siding is usually milled into special siding patterns with interlocking grooves that overlap the edges of the bevels. However, ordinary plank siding is not patterned in such a fashion. Nor are the pieces used in regular board-and-batten or other combinations of planking.

A number of popular uses of siding, both milled and straight-edged, are described below.

Straight-edged board planks can be applied as siding in a number of ways, with three designs the most commonly used: board-and-batten, board-on-board, and clapboard (unmilled).

Board-and-batten is applied vertically, with two sizes of strips. The boards are wide, the battens thin, with the boards fastened to the sheathing, and the battens covering the joints between the boards. Face-nail each underboard once with 8d nails; space ½ inch apart. Face-nail each bat-

ten (or overboard) with 8d or 10d nails with 1 inch overlap.

Board-on-board is applied vertically, with each board the same size. Face-nail underboards once with 8d nails, and face-nail overboards twice with 10d nails.

Clapboard is applied horizontally, with straight-edged boards. Face-nail each board once with 10d nails, 1 inch from overlapping edge.

Straight-edged planks come in 1-inch thicknesses and in widths of 4, 6, 8, 10, and 12 inches. Actual size of a 1 by 4 is 3½ inches wide; 1 by 6, 5½; 1 by 8, 7¼; 1 by 10, 9¼; and 1 by 12, 11¼.

Channel rustic, also called **board and gap,** can be applied either vertically or horizontally. When in place, the siding will provide a ½-inch lap and a 1¼-inch channel. Face-nail each board once with 8d nails for 6-inch widths. For wider, use two nails per board.

Drop siding is applied horizontally, and provides a slight bevel at the top of each board to accentuate the joint. It comes in tongue-and-groove patterns or shiplap. Blind-nail tongue-and-groove with 6d finish nails; face-nail shiplap with 8d siding nails.

Bevel, also called **Colonial, Bungalow, Dolly Varden,** and so on, is available plain or with rabbeted edge. It is intended for horizontal application. Face-nail each plain board once. With the rabbeted bevel, face-nail each board 1 inch from the lower edge with 8d nails. Bevel siding comes in 4-, 6-, 8-, 10-, and 12-inch widths, with actual widths the same as plank boards.

Plain *tongue-and-groove* can be applied vertically, horizontally, or diagonally, with mixed widths in random applications. Blind-nail tongue-and-groove with 8d nails, or face-nail with siding nails. Blind-nail 4- and 6-inch widths with one nail. Face-nail wider boards with 2 nails to each stud.

Actual face widths of tongue-and-groove are: 1 by 4, 3⅛ inches; 1 by 6, 5⅛; 1 by 8, 6⅞; 1 by 10, 8⅞; and 1 by 12, 10⅞.

Log cabin comes in several different forms. There is an actual round log that can be fitted to another round log, the whole interlocked into a cabin or house wall. Such logs can be purchased from manufacturers who precut each log to its desired length for immediate assembling at the

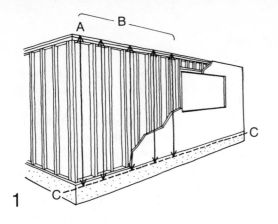

1

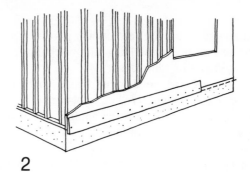

2

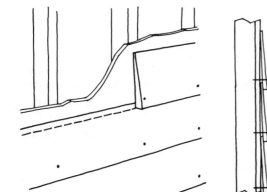

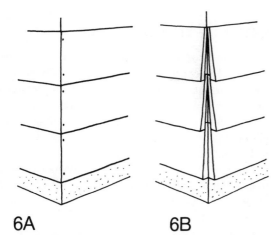

3

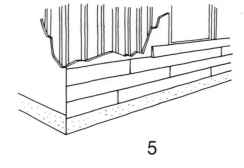

4

5

6A   6B   7

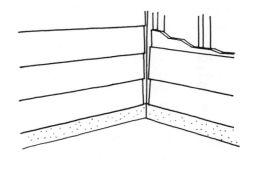

Fig. 149. Drawings show how to apply bevel siding in detailed step-by-step instructions. Measure from the top of the top plate to the point below the sill, (usually the foundation), and mark, as shown in (1-A). Repeat the step every 2 feet (1-B, etc.). Snap a chalk line horizontally along the marks (1-C) to provide for the alignment of the first course. Position butt edge of siding strip along chalk line (2). Nail first course to sill plate at points below each stud. At finish of first course, snap a chalk line 1 inch below the top of the course (3). Nail second course so nails clear the thin upper edge of the bottom course by ⅛ inch (4). (5) Other courses going up. Make sure vertical butt joints are staggered along the side wall and that all joints fall on studs. Outside corners can be mitered or capped with metal corners as in (6-A) and (6-B). (7) Shows how an inside corner can be cut to length prior to application and butted against a square wood corner strip 1⅛ by 1⅛ inches in size.

building site. These logs are usually part of a prefabricated or precut building kit which must be purchased as a building unit.

For the average wood-frame house, there is a type of log cabin veneer that is between 1 inch and 2 inches thick and can be applied either vertically or horizontally outside an ordinary sheathing surface. These pieces come in widths of 10 to 12 inches. Face-nail with 10d or 8d siding nails, depending on the instructions of the suppliers.

Shiplap can be applied horizontally or vertically. Face-nail once with 8d nails for 6-inch widths, and twice for wider types.

Actual face widths of shiplap are: 1 by 6, 5⅛ inches; 1 by 8, 6⅞; 1 by 10, 8⅞; and 1 by 12, 10⅞.

TYPES OF WOOD

Redwood and cedar are the best solid-board siding woods because they are easy to work with, are free from warp, and have a natural resistance to decay. They also take a paint finish very well.

Other woods include sugar pine, western white pine, eastern white pine, cypress, yellow poplar, ponderosa pine, western hemlock, and spruce, southern yellow pine, western larch, and Douglas fir.

Shop for vertical-grain board primarily of heartwood—the center part of the log that is more durable. They should be graded "clear" or "select." Flat-grain surfaces warp more easily than vertical-grain surfaces. Boards with tight knots can be used for paint surfacing, provided they are treated before installation.

See Chapter Five on shopping for lumber and for hints on picking out the best siding for your house. Most building-supply houses carry solid-board siding that is preprimed for paint or pretreated with water repellent or sealer.

Because of its essentially simpler milling, plank siding is usually cheaper per board foot than tongue-and-groove or shiplap. And, since there is no overlap, you will get more coverage from a series of 1 by 12s than from 1 by 12 tongue-and-groove boards.

However, tongue-and-groove and shiplap strips give more protection against the elements because of the tight joints between members. Also,

the milled joints tend to make the boards fit together better and straighter as they are nailed in place.

Most wood siding is soft and easily damaged. It is a good idea to handle boards with care when they are delivered to the building site. Stack the strips off the ground and keep the pile under cover from rain and moisture.

In applying siding, do not let adjoining strips end on the same stud. Also, in nailing pieces above and below windows and above doors, do not let the strips end inside window or door zones. Be sure the ends fit in tightly at doors and windows; it is in these areas that the most weather damage occurs in siding. Fit boards tightly at the corners of the house, too.

To prevent splits during application, use blunted nails on all board siding. You can buy blunted nails, or you can blunt them yourself with a hammer. Do not drive nails too near the edges of siding pieces. If the wood is very soft, and if you must nail near an edge, predrill all nail holes shank size to be on the safe side.

EXTERIOR PLYWOOD SIDING

Plywood's strong construction, its large panel sizes of 4 by 8, 9, and 10 feet, and its wide variety of surface textures, make it a siding material that saves time and labor in every way. It is easily and quickly put up, provides extra strength, and eliminates the need for cross bracing or structural wall sheathing.

Textured varieties of exterior plywood include striated, fine line, brushed, kerfed, rough sawed, reverse board-and-batten, channel groove, and texture 1-11.

Large plywood modules can be attached either vertically or horizontally. If you nail them on horizontally, stagger the vertical edge joints. Attach the long edges into fire stops or other nailing blocks in the middle of the wall.

For exterior plywood installation, nail the edges every 6 inches and the inside supports every 12 inches. Use 6d nails for ⅜-inch or ½-inch panels, and 8d nails for ⅝-inch panels.

Be sure to leave 1/16-inch space between all panel ends and edge joints for expansion under extreme weather conditions.

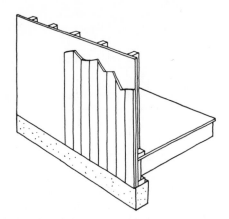

Fig. 150. Plywood for exterior siding use comes in a wide variety of surfaces and textures. Shown is a type of plywood exterior that can be applied to non-structural sheathing. No building paper or diagonal wall bracing is required with plywood-panel siding. Some plywood techniques are designed for horizontal application.

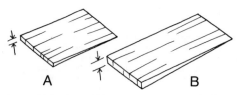

Fig. 151. Shingles and shakes are sometimes indistinguishable from one another. Usually a shingle is milled and finished off with machinery, and a genuine shake is split by hand. Shingle (A) is smaller than shake (B) and thinner. Average shingle is about ¼ inch thick at its bottom, or thickest portion, and about 18 inches long. Shake is about ½ inch thick at its bottom portion, and about 24 inches long. Shakes are more expensive than shingles, but last longer.

SHINGLES AND SHAKES

Wood shingles can be used effectively not only for shed roof surfaces, but for exterior walls as well. They can also be used on interior surfaces for special accents. A shingle differs from a shake in that it is processed and milled more closely, and is smaller. A shake is usually handsplit, and is more expensive than a shingle. Siding shingles vary according to grade, size, texture, and cut.

Quality is graded by numbers ranging from 1 to 4, with Number 1 of 100 percent vertical-grain heartwood; Number 2 of minimal flat grain and sapwood; Number 3 of utility grade; and Number 4 for undercoursing and interior walls.

Shingles come in common lengths of 16, 18, and 24 inches; widths from 3 to 14 inches.

If you want to avoid attaching each individual shingle to the side of the house, you can get large commercially prepared panels of shingles bonded to a backing of ½-inch sheathing-grade plywood. The panels contain a row of 18-inch shingles attached to an 8-foot strip of plywood. You attach the 8-foot strip to the wall surface for a normal course.

Shakes are all Number 1 grade. Thicknesses vary from ⅜ inch to 1¼ inches; lengths 15, 18, 24, and 32 inches. Widths are random.

Shingles for siding should repell moisture and cold, but they don't have to be as tough or as durable as roofing shingles which have to withstand severe precipitation and hot sunshine. Applied as decoration to inside walls, shingles or shakes will get much less wear and can be that much less sturdy.

Solid sheathing or open-board sheathing can be used as underlay for shingle or shake application. It is preferable, however, to use solid plywood sheathing to give the shingles or shakes more structural support.

Nail a layer of 30-pound building felt with tar-paper tacks over the sheathing before beginning to nail on the shingles.

You can apply both shingles and shakes in either **single** or **double courses.** A course is a row of shingles or shakes placed side by side along the length of a wall or roof. Each course should overlap about one-half the course directly below it. This means that there will be three layers of wood at all points—two of shingle and one of sheathing.

A double course is simply two layers of shingles, one over the other. You use low-quality shingles for the underlayer, and high-quality shingles for the overlayer. The underlayer is not visible to the naked eye, inasmuch as it is laid ½ inch out of sight under the overlayer.

Exposure refers to the part of the shingle or shake that is open to the weather. The amount of exposure depends on the length of the shingle or shake, and is about half or a little more of the length of the shingle, at least on a single course.

Proper exposure for siding shingles is indicated in the table below, which gives approximate measurements for both single and double courses.

Shingle Width	Single Course Exposure	Double Course Exposure
16 inches	7½ inches	12 inches
18 inches	8½ inches	14 inches
24 inches	11½ inches	20 inches
32 inches	15 inches	—

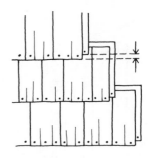

Fig. 152. Picture shows how a double course of shingling is laid. Note that outer course runs just about ½ inch lower than inner course. Both courses are nailed into sheathing with 30-pound building paper over sheathing.

It is easy to estimate the number of courses needed for a house wall by using the figures in the box. Measure the wall from bottom to top in inches and divide by the amount of exposure.

HOW TO APPLY SHINGLES AND SHAKES

Shingles and shakes are always laid in courses starting at the bottom of the wall and moving upward. Each higher course overlaps the lower by a specified number of inches.

The number of inches of overlap is the difference between the length of the shingle and the amount of exposure. A 16-inch shingle in a single course has an exposure of 7½ inches. The overlap would be 8½ inches.

Before beginning, mark the lines of the courses on the corners of the outside wall. Each course line will show where the bottom or butt end of the shingle—the wide end—will be placed. The amount of exposure can be varied slightly in order to avoid breaking shingles under windows or above doors. Try to keep the shingles running evenly above and below windows.

When the course lines are arranged, snap a chalk line along the lowest course just above the foundation line. Attach a line of shingles along the first course. Then nail a *second* course over the first one. The first course of a single application should always be a double one.

Shingles and shakes can be nailed to an old wooden wall, to an old shingle surface, or to ply-

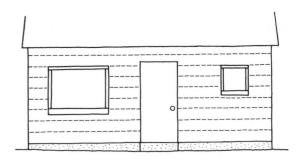

Fig. 153. Before applying siding shingles to house wall, measure distance from bottom to top, and select desired amount of exposure. Try to keep lines of shingles from breaking just above or just below windows and door frames. Snap chalk lines along bottom run, and lay courses. On right window, insert small bottoms under sash, or apply horizontal strip of wood the same kind and color as shingles.

wood sheathing. It is possible even to shingle over a stucco or plaster wall. Simply apply furring strips horizontally, spacing the strips so they can be used as a nailing base for each single course. Place the strips so their centers run an inch above each course line.

To proceed, place each shingle with the butt end exactly on the course line. Then blind-nail it an inch above the chalk line for the next course. The butts of the next course will cover the nail by an inch. Nail shingles ⅛ to ¼ inch apart.

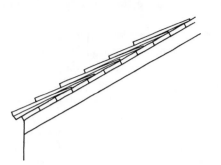

Fig. 154. Schematic drawing shows how shingles are laid on roof. Note that first course at bottom near eave is double course. Shingles are attached to sheathing as shown, with building paper between. "Run" is distance from end of rafter to top of roof. "Rise" is distance from bottom of rafter to top of roof. "Exposure" is amount of shingle showing to the eye and "exposed" to the elements.

Two nails should be used for shingles up to 8 inches wide, each fastened ¾ inch from the edges. Wider shingles need a third nail in the middle. Do not smash the wood with the hammer.

For each course after the first double course, nail a long guide board just below the chalk line. Push each new shingle butt tightly against the guide and nail it in place. Do not allow vertical joints between shingles to align with the course directly below.

At the inside corners of two walls, fasten square vertical wood strips for the shingles to butt against. On outside corners, miter the end shingles or alternate them as overlaps.

For applying double course shingles and shakes, nail the inner course about ½ inch above the course line, so it will be out of sight when the second shingle is fastened on. Begin the bottom course with three rows of shingles, one extra undercourse on top of the regular undercourse and overcourse.

First fasten each undercourse shingle once with nails or staples. Then fasten the finish shingle on top of it about ½ inch lower, fastening it 2 inches above the next shingle's butt edge, ¾ inch in from each side.

Proceed as above.

REPAIRING OR REPLACING BOARD SIDING

Small cracks, warped boards, or open joints in siding should be attended to promptly with repair or replacement of the siding strip. These flaws not only cause paint to peel, but siding to rot, ruining complete sections sometimes.

Minor repairs to siding can be made by filling in the cracks and joints with a good grade of caulking compound or white-lead putty. Pack the fill in tightly. Nail the board down along the edges if it has worked or warped loose.

Drill pilot holes for the nails to prevent the splitting of dried old wood. Countersink the nails below the surface of the siding, and fill the holes with putty. Let the putty and caulking dry for a day or two before touching up the surface with exterior house paint.

When a length of siding board is badly cracked, severely weathered, or rotted, it must be replaced with a new board. For the neatest replacement, remove the full length and put in a new board. If removal of the entire length is unfeasible, cut the defective portion out and replace it with a new piece cut to fit.

Cutting out a segment of a board is not always easy. Since many siding boards overlap, the board to be replaced will be held along its top edge by nails driven through the lower edge of the board above.

Drive wooden wedges under the damaged board to pry it out slightly from the board below, or pull it out an inch or so with a pry bar. The object is to get it away from the boards above and below for cutting.

Tape blocks of wood into position to protect adjacent siding. Then, with a handsaw or an ordinary crosscut, saw through the damaged board.

After making cuts at both sides of the rotted section, pry up the board above so you can get at the remains of the rotten board underneath.

Split away the old cracked board in sections by prying and chopping with a chisel. Avoid driving the chisel through the board into the building paper beneath.

If a narrow strip of broken clapboard remains under the edge of the top board, cut the nails that hold it by forcing a hacksaw blade in between the

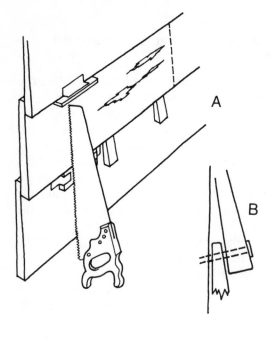

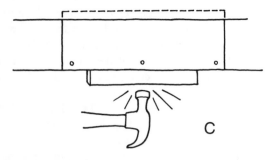

Fig. 155. Pictures show one way of replacing section of siding. (A) Shows how to saw bad piece using handsaw or backsaw. Tape blocks of wood on upper and lower boards to protect them from the saw marks. Insert wooden wedges under broken portion of clapboard. Remove sawed-out piece. If some scrap remains under upper board, cut the nail in two places as shown in side view (B). Cut new board to size, and insert it into the old spot (C). Use block of wood and hammer to drive it back in under the upper board. Nail in place.

top board and the strip of the bad board. With the nails cut, pry out the strip with a narrow chisel or screwdriver.

After the old piece has been removed, check the building paper for tears or cuts. If there are any, seal them with roofing cement. If large gaps are noticeable in the paper, nail a new sheet over

the entire area, sealing the edges with roofing cement.

Cut a new length of matching clapboard to replace the damaged section. Install it by tapping upward from the bottom, using a block of wood to avoid damaging or denting the edge.

Line up the lower edge of the new board at each end, and fasten it in place by face-nailing along the bottom edge about 1 inch up. Face-nail the lower part of the board above the replacement to catch the top edge underneath.

Countersink the nailheads, putty the holes, and paint the board with exterior primer as soon as possible. After the first coat is dry, seal the end joints where the new board meets the old with caulking compound.

Fill the nail holes with putty. Apply a final coat of paint to blend the patch in with the surrounding wall area.

REPAIRING OR REPLACING SHINGLES

Inspect wood and asbestos shingles covering exterior walls at least once a year to make certain that none of them are split or missing. Also check each one for damage that might necessitate replacement.

Shingles tend to sag or bulge when the nails holding them work loose. Hammer the old nails back in, taking care not to break the shingle's surface.

If the nails still fail to hold, drive in an extra nail an inch or two away from the old one. Use rustproof threaded or ringed nails. They grip better and are less likely to work loose in the future.

Drill a pilot hole before driving nails through asbestos shingles to avoid splitting them. They are very brittle. Wood shingles generally do not need predrilling, but if they are old and dry, or if a nail must be driven close to the edge, use a pilot hole. In order to prevent damage to the shingle, be sure not to hammer the nail in too far.

When a shingle is split with both halves still in place, you won't need to replace it. Slide a sheet of building paper under the split and drive nails in on both sides near the butt of the shingle.

If the shingle is cracked in more than two places, or if a shingle has been badly smashed, replace it. Since all exterior wall shingles, both wood and asbestos, are installed with the lower half overlapping the upper part of the one below it, you will have to remove the nails holding the shingle on top in order to free the upper edge of the broken one. The basic technique of removing and replacing shingles and shakes is the same as with siding (see above).

Wood shingling is occasionally double-coursed; cut the nail between the broken shingle and the lower course of the next double course. Double-coursing is sometimes simulated by a false run of undercoursing, usually of fiberboard strips or perhaps narrow pieces of wood like furring, to bulk out the wood shingle and give the shadow effect that double-coursing provides.

Cut the nails just above the broken shingle as above.

After the nails are sawed, pull the shingle forward at the bottom and pry out the nails along the lower edge. Slide the shingle cut from the bottom so that a new one can be inserted in its place.

CHAPTER TWELVE

Roofs

Working on a roof is an extremely dangerous job. It may look like sport—mountain climbing always does—and it may seem that you can keep your footing easily, but do not be deceived. A 45-degree slant can be a very tough slope to negotiate without the help of ropes or cleats nailed onto the roof—especially when you're carrying tools and roofing material as well.

If in doubt, don't climb up on a peaked roof to do a job or even to look for defective shingles. Hire a professional to do it for you. If you feel that you can acquire the habits of a mountain goat, do so, but you must realize you are running a risk that is formidable.

HOW TO WORK SAFELY ON A ROOF

Whatever you do, never go up on any kind of a roof without wearing shoes with rubber soles and heels. Stiff soles and leather heels are simply foolhardy gear to wear on any slope.

In addition, there are several ways to work safely at a height. One of them is so simple it might appear to be a joke, but it is not. Secure yourself to the end of a rope, and tie the rope either to the chimney, vent pipe, or if there is nothing strong enough, to something on the ground *on the other side of the house peak.*

In moving around on the end of a rope, keep retying the cord wherever you are to take up the slack. Be sure to keep your feet planted as flat-footed as possible against the surface of the roof. It is the tilted-in shoe that begins to slip, not the one planted firmly on the surface for maximum purchase.

You can always tie two ladders together at the peak of a roof, letting each ladder stretch down one slope. From one of these ladders you can reach almost any point of the roof by sliding the double ladders back and forth along the peak.

Professional roofers frequently nail cleats along the surface of the roof as they proceed up. If you use this method be sure to secure the cleats firmly into the sheathing through the roofing surface. The problem with this practice is that the removal of the cleats leaves holes in the roofing.

On a roof pay close attention to your work and *do not look down too much.*

STRUCTURE OF A ROOF

The roof of a wood-frame house, flat or peaked, is constructed in the same fashion as a wall or floor. The outside surfacing is supported by a series of skeletal framing members called rafters. The rafter is equivalent to a wall's stud and a floor's joist.

Roof rafters are usually framing timbers 2 inches by 6, 8, 10, and 12 inches, depending on the size of the roof span and on the weight of the roofing material. Building codes also determine the size of roof rafters.

To the roof rafters on the outer, or top, side, is attached a cover of sheathing which, in turn, supports the outer surfacing, the "roof" you see from outside. Sheathing comes either in strips of tongue-and-groove or shiplap, or in plywood modules. Building paper is attached to the sheathing by means of some kind of adhesive or fastener —usually roofing cement, of which there are many formulations now in use.

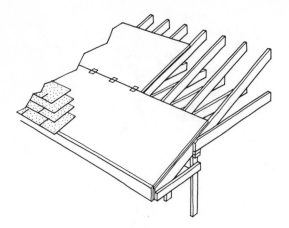

Fig. 156. Illustration shows the anatomy of a pitched roof on a typical wood-frame house.

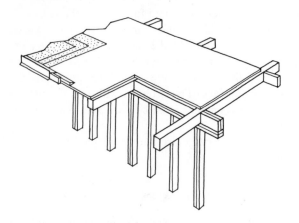

Fig. 157. Illustration shows the structure of a typical flat roof on a wood-frame house.

In most cases the rafters of a flat roof are spliced together in the middle over a girder that runs parallel to the load-bearing walls. A peaked roof meets in the middle at the ridgepole, which acts almost as a support girder. The ceiling rafters of a peaked roof generally meet and are spliced together over a girder or wall running parallel to the load-bearing walls.

In a flat-roofed house, the ceiling can be attached to the bottom of the roof rafters. However, in most more elaborate structures, there is usually a space between roof and ceiling where air is allowed to circulate for ventilation. The rafters for ceiling surfacing are sometimes smaller than support rafters for the roof.

A peaked-roof house is usually surfaced with sheathing and a number of different types of surfacing: asbestos shingles, asphalt shingles, wood shingles and shakes, slate, and so on (see Fig. 154). Most of these surfaces require building paper between the sheathing and the outer surface. The two most common types of surfacing for peaked-roof houses are wood shingles/shakes, and asphalt shingles.

In all cases, the surfacing of a peaked-roof house is applied to sheathing in either strips or in plywood modules. Felt building paper—30-pound—is usually tacked to the sheathing with roofing tacks before the shingles are applied. Wooden shingles and shakes are attached by driving shingle nails into the sheathing through the paper. Asphalt shingles are attached by driving shingle nails into sheathing after the paper is treated with a special waterproof roofing adhesive or cement.

The top of a flat-roofed house is usually surfaced with sheathing and building paper. Strips of roofing paper covering a flat roof overlap a number of inches, usually specified by building codes. Tin and metal can be used to cover a flat-roofed house, too, but neither is common in homes.

Wood and asphalt shingles are laid in runs, beginning at the bottom of the peaked-roof house at the eaves and moving up toward the peak. At the hip of the roof, special shingles are attached which act to seal the peak joint. The shingles shed water directly into the gutters and downspouts, which are attached either to the top of the wall at the juncture with the roof or to the roof sheathing at the eave.

The technique of shingling is similar to that described in Chapter Eleven on exterior siding.

Where peaks, gables, and valleys meet there is maximum danger of leaks. Before shingles are applied, these roof joints are covered with wide strips of metal called flashing. The flashing is a gutter that directs water downward toward the drains. Shingles are applied over the flashing, covering it completely.

At every roof joint, including where chimneys and ventpipes meet shingling, flashing is supplied as a protective cover between the sheathing and the shingles. In this way, every roof joint is covered not only by sheathing and shingles, but by a layer of metal flashing as well.

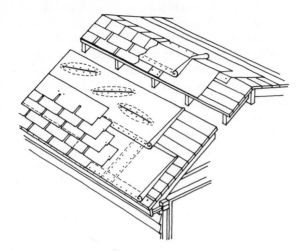

Fig. 158. Diagram shows a typical peaked roof where asphalt tiles have been installed over old roof surface.

To repeat the original warning: roofwork is dangerous, and construction jobs should be handled only by professionals. However, repairwork —especially that dealing with the patching up of leaks—is within the province of the amateur carpenter, and doing this yourself can save money and prevent a lot of damage by rampant moisture within the house. But be careful on the roof, and pay attention to all safety rules!

HOW TO SPOT A LEAK

There are two kinds of roof leaks—the obvious and the hidden. The obvious leak is caused by a fully visible piece of roofing material like a broken shingle, a bent piece of flashing, or a torn asphalt shingle. The hidden leak is caused by some flaw you cannot see. It is much more tricky to find, although the actual repair may be the same in each case.

The hidden leak can be a very aggravating problem. Suddenly water appears on a ceiling downstairs. Upon investigation, the roof checks out perfectly, with nothing to indicate where the water is entering. It can take a great deal of detective work to ferret out the leak.

Begin at the appearance of the leak and work backward. Remember that water always flows— at an angle, straight down, spreading outward,

and even in some cases by absorption going even slightly upward, to spill over and go down again.

A leak in a peaked roof may start near the top of the peak, and not appear again until it comes out at a ceiling fixture in a downstairs room. Water may enter the roof, travel along an insulation batt, enter a wall, run along a ceiling joist, transfer to an electric wire, and drip down through a fixture.

The best time to check out any leak is during a rainstorm, when it first appears. Go upstairs and climb into the crawl space under the roof rafters. Using as much work light as you can, inspect the entire section above the leak. If one of the rafters is wet, trace it until you find the source. Mark the spot with a pencil or crayon.

When the storm lets up, measure the distance from the mark inside the house up to the peak of the roof. Climb onto the roof. To get to the right rafter, simply count the rafters inside, multiply by 16—or 24 if the rafters are 2 feet on center— and measure from the edge of the roof in inches. Measure the same distance down from the peak as inside and mark the leak.

At the suspected site, inspect the shingles there and the shingles immediately adjoining. In most cases you will find there is a bad joint, a split piece, or a hole in a blind spot where the rows of shingles join underneath.

Replace the material as explained below.

Once the piece is repaired or replaced, make your own storm by watering the roof of the house with a garden hose to check out the leak to be sure it has been sealed tight.

If you have no crawl space, or if the roof rafters are sealed by insulation and covering, you may have to open up the board at one of the joints. Push your fingers inside to test for water. It may be dripping in through a leak at the top of the slope and running down the top part of the insulation batt. Trace it up to the highest point and mark that spot, and proceed as above.

HOW TO REPLACE A SHINGLE OR SHAKE

Although a wood shingle on a roof may endure indefinitely, some may develop splits from shocks or blows to the surface. When a shingle splits, it may allow moisture to get in behind it and cause it to rot.

A shingle cannot easily be repaired; it must usually be replaced.

To replace a damaged shingle, it is necessary to remove the shingle nails that hold it to the sheathing and the shingle nails that hold the shingle on the next course up—since that nail also holds the top of the damaged shingle.

Use a hacksaw blade and cut up through the space between the shingles. Then remove the nailheads. Saw through the shingle nails of the damaged shingle first, then saw through the nails attaching the shingle in the course above. Remove all nailheads. You can probably work loose the damaged shingle, pulling it down and away from the sawed-through nails.

It is also possible to use a cold chisel to break apart the damaged shingle, reducing it to thin pulverized splinters without sawing off nails. However, once the splinters are taken away, it is necessary to remove the nails attaching the shingle in the next course up to enable you to put in a new shingle. With a block of wood, pound the shingle butt lightly and let the nails ease out. With the claw of the hammer you can usually remove them without trouble.

Prepare a shingle by cutting it to the same size as the damaged one. Push it up into the slot where the old one was positioned. When it is in place, nail it, using rustproof nails. Pound the nails in an inch or so away from the old nail site. Renail the shingle on the next run up.

Since there frequently is need to replace a shingle on a roof or on a house's siding, it is a good idea for any homeowner to have a pack of extra shingles in store for use at any time. Colors may be matched more easily in that fashion, and types of shingles and shakes will not be obviously out of harmony with the whole.

HOW TO REPLACE AN ASPHALT SHINGLE

Damage to an asphalt shingle can sometimes be repaired by a simple application of roofing cement. A small piece can be patched up by adding a shingle segment of similar size over the tear or flaw with roofing cement.

If the hole is too big, or if the damage is too extensive—as it might be in a situation when a high wind has ripped a whole section loose—it is necessary to replace the entire length of shingle.

Since asphalt shingles come in combinations of three, it is usually easier to replace the whole section rather than part of one.

To remove a section, fold up the shingles in the course above the damaged section. The nailheads holding in the damaged section will be revealed. Remove the nails with a claw hammer, digging into the torn asphalt in order to get the proper purchase.

If for some reason this proves to be unfeasible, use a hacksaw blade in the manner described in

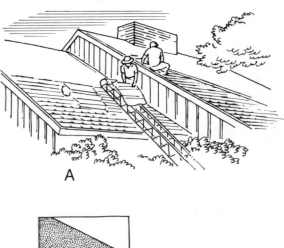

A

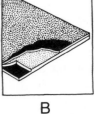

B

Fig. 159. Asphalt shingles can be applied over most existing roofs. (A) Shows them being installed over old wood shingles. Let a professional roofer do this rather tricky job. It'll save time and get you a guarantee. (B) The composition of an asphalt shingle: back coating prevents weathering from underneath; felt base is saturated and waterproofed with asphalt; heavy asphalt coating aids longevity; top layer of mineral granules provides fire resistance and color.

the section above, sawing through the nails on the underside of the damaged shingle. Take care not to damage the building paper beneath.

When the defective section is removed, inspect the building paper and apply adhesive and sealer to patch up any holes or tears in the paper.

Then place a new section of shingling where the old one was, nailing quickly in place, and placing roofing cement on top of each nailhead.

Let the shingles in the course above cover the nails and cement. If necessary, apply more roofing cement to hold the upper course in place.

Since asphalt shingle colors, textures, and designs are difficult to match from year to year, it is always wise for the homeowner to keep a separate packet of them available for emergency repairs.

SHINGLING A ROOF

Shingling a roof is a job for a professional, not a home handyman. The techniques used for applying roofing shingles—wood and asphalt—are quite similar to those used in applying shingles to siding. However, working on a rooftop is much more dangerous than working on the side of a house. Do not try to do the job unless you are a skilled technician.

HOW TO REPAIR A FLAT ROOF

A hole in a flat roof can be repaired by swabbing the affected area with roofing cement.

If the hole is very large, and if it looks as if a simple patching job won't plug up the hole sufficiently, you should replace the affected part with a new square or rectangle around the damaged area.

To repair a large break, lay a sheet of new building paper so that it completely covers the break in the paper below. With a utility knife, cut through both the new paper and the old paper simultaneously. When the cut is made, peel up both the rectangles or squares.

Throw away the old damaged paper.

If segments of the cutout rectangle cling to the roof, remove them by cutting away from the adhesive, or simply pull them up.

Clean the sheathing below the roofing paper until it is spotless and dry. Swab the surface with roofing cement, and place the new rectangle exactly in the cutout spot. Nail the new piece in place with roofing nails spaced closely together along the perimeter of the patch piece.

Cover the nailheads with roofing cement, and lay a bead of roofing cement along the joint between the old roof and the new patch.